NanoScience and Technology

NanoScience and Technology

The series NanoScience and Technology is focused on the fascinating nano-world, mesoscopic physics, analysis with atomic resolution, nano and quantum-effect devices, nano-mechanics and atomic-scale processes. All the basic aspects and technology-oriented developments in this emerging discipline are covered by comprehensive and timely books. The series constitutes a survey of the relevant special topics, which are presented by leading experts in the field. These books will appeal to researchers, engineers, and advanced students.

Please view available titles in *NanoScience and Technology* on series homepage http://www.springer.com/series/3705/

Changxin Chen
Yafei Zhang

Nanowelded Carbon Nanotubes

From Field-Effect Transistors to Solar Microcells

With 69 Figures

Springer

Dr. Changxin Chen
Dr. Yafei Zhang
Shanghai Jiao Tong University, Research Institute of Micro/Nano Science and Technology
800 Dong Chuan Rd., Shanghai 200240, P.R. China
E-mail: chen.c.x@sjtu.edu.cn, yfzhang@sjtu.edu.cn

NanoScience and Technology ISSN 1434-4904
ISBN 978-3-642-01498-7 e-ISBN 978-3-642-01499-4
DOI 10.1007/978-3-642-01499-4
Springer Heidelberg Dordrecht London New York

Library of Congress Control Number: 2009929207

Printed on acid-free paper

Springer is part of Springer Science+Business Media (www.springer.com)

Foreword

The discovery of multiwalled carbon nanotubes (CNTs) in 1991 and the subsequent discovery of single-walled CNTs in 1993 have led to a worldwide excitement to explore their fundamental properties and potential device applications. A single-walled CNT is structurally a sheet of graphene rolled into a seamless tube, which possesses a diameter of the order of a nanometer but a length thousands of times greater. The large aspect ratio and small cross section size make it nearly an ideal, quasi-one-dimensional system, which has provided a concrete context for chemists, physicists, and engineers to collaboratively work together in the field of nanoscience and nanotechnology. As a result of such efforts in the last two decades, superior electrical, optical, and mechanical properties of CNTs have been theoretically predicted and experimentally demonstrated. The unique material properties of CNTs have made it interesting for a variety of applications. For example, depending on the orientation of its graphene lattice relative to the nanotube axis, the CNT can be either metallic or semiconducting. This property makes the material interesting for developing not only nanoscale semiconductor devices but also a new interconnect technology to compete with the state-of-the art copper interconnect technology. The property has also imposed yet-solved challenges in the field. For example, one of the major challenges that holds CNTs back from electronic application is how to produce pure all-semiconducting CNTs, based on which a device that can be effectively turned off can be built. In addition, the bandgap of a semiconducting single-walled CNT, which stems from electronic confinement in the circumferential direction, is inversely proportional to its diameter. As the characteristics of a semiconductor device strongly depend on the bandgap, this property offers a new dimension of freedom to design and tune the properties of CNT electronic and optoelectronic devices. On the other hand, it also imposes the challenges to control the variability between devices, or even between different tubes in the same device. Introducing the promising CNT material into device technologies will certainly open new doors. It, however, still demands significant research and development efforts to be devoted.

The advances on synthesis, material quality, and device performance of CNTs have sparked strong interest to turn the excellent material properties into new device technologies. As an example of exploration on CNT-based devices, CNT field-effect transistors (FETs) have been extensively investigated in the last decade and regarded as one of the most promising candidates to sustain the continuous growth of electronics industry in the postsilicon CMOS era. In a CNT, low bias transport can be nearly ballistic, with a scattering mean free path in the micrometer scale. Deposition of high-κ gate insulators does not degrade the carrier mobility because of an absence of dangling bonds. The conduction and valence bands are symmetric, which is advantageous to complementary applications. The bandstructure is direct, which enables optical emission and the potential to integrate devices for computing and devices for communication on the same nanoscale material platform. Since the first demonstration of the CNTFET in 1998 by Dekker's group at Delft University of Technology and Avouris's group at IBM T. J. Watson Research Center, the progress in understanding its device physics and improving device performance has been substantial. An individual channel CNTFET with its intrinsic performance surpassing state-of-the-art silicon CMOS has already been demonstrated by Dai's group at Stanford University.

Although the progress has been substantial, an industrialized CNT device technology still does not exist, and major barriers remain. For example, despite the excellent carrier transport property, the CNTFET does not deliver a large enough current that is necessary for driving interconnects and logic gates in an integrated circuit, due to its nanometer-wide channel. The capability to put multiple, uniform, and closely packed semiconducting CNTs in parallel with the transistor channel will solve the problem, but the task is challenging because the synthesized CNTs come as a mixture of metallic and semiconducting species, with a statistical distribution of positions, orientations, and diameters. Furthermore, just as true for other semiconductors, contacts play an important role in CNT devices. The Schottky barrier between a metal contact and a CNT is known to deteriorate the transistor and optoelectronic device performance. Chemical doping of a CNT, which consists of a rolled-up monolayer of carbon atoms, is more difficult and does work as well as that of silicon for producing an Ohmic contact. Fortunately, an Ohmic contact to an even intrinsic CNT can be achieved by correctly choosing a contact metal material and a right range of CNT diameter, as experimentally demonstrated by the Stanford group. The pursuit of methods to improve reliability and reproducibility of the contacts, which is especially important to circuit integration where high device yield and low variability are crucial, is still underway.

The book, Nanowelded Carbon Nanotubes: From Field-Effect Transistors to Solar Microcells, by Changxin Chen and Yafei Zhang, presents important progresses toward CNT device technologies. Especially interesting topics include the development of and the progress on the nanowelding techniques for CNT contacts and the alignment techniques for a multiple CNT channel, and their successful applications to the CNT field-effect transistors and CNT solar microcells. Chapter 1 reviews the structure and electronic and transport properties of CNTs. Chapter 2 provides an overview of CNT production by synthesis and purification. The issues of aligning CNTs and contacting CNTs to metal are addressed in Chaps. 3 and 4 respectively. The last two chapters of the book present device applications. Chapter 5 describes a nanowelded multichannel CNTFET enabled by the CNT alignment technique for a parallel channel and the nanowelding technique for the metal–CNT contact. The photovoltaic effect and energy conversion by CNT solar microcells are discussed in the last chapter. As the CNT provides a concrete context for studying the fabrication, physics, and performance of new devices enabled by a new one-dimensional material, the knowledge should also provide invaluable lessons in the pursuit of new electronic and energy conversion devices enabled by one-dimensional nanomaterials in general.

Gainesville, FL, January 2008 *Jing Guo*

Preface

Over the past one and a half decades, carbon nanotubes (CNTs) have evolved into one of the most intensively studied materials. For their unique one-dimensional nanostructure and excellent physical, chemical, and mechanical properties, CNTs are considered as an ideal building block of next-generation electronic and opto-electronic devices and circuits. The research effort devoted by both academic and industrial community has induced a great advance of the science and technology for CNTs in all fields including its synthesis, property, device application etc. It is becoming evident that CNTs (specifically single-walled CNTs) have an ability to replace silicon in electronic devices that dominate our present information/data-driven world. Despite the extremely promising prospect, many critical tasks and challenges still lie ahead to realize the industrial application of CNT devices with a small size, high speed, and low-power dissipation, including the control of CNT quality and placement, the development of suitable fabrication processes, the integration of CNTs into circuit architectures etc. Thereinto, the contact between CNTs and external structures plays a crucial role, which will significantly influence not only the device performance but also the device reliability and stability. The simple, low capital intensive, and scalable processes for the formation of robust contacts are a premise to build high-performance CNT-based devices and systems for future large-scale applications. With this spirit, a concern on the contact formation and optimization and its device applications has been given in this monograph.

A number of books have been published in the past decade outlining a variety of topics of current interest related to the research of CNTs. These books deal with topics ranging from synthesis methods of CNTs to unique physical and mechanical properties of CNTs and their possible real-world commercial applications in the future. On a cumulative basis, this book places an emphasis on the introduction and discussion of a novel ultrasonic nanowelding technology and its device applications for CNTs. With this technology, the CNTs can be bonded onto metal electrodes to form firm, low-resistance CNT/metal contacts and can significantly improve the device performance. The CNT field-effect transistors and solar micro-cells with nanowelded contacts are also looked at in this book. In order to present the readers with a more correlative learning on CNTs, the book also covers the structure and electronic property of CNTs, synthesis and purification of CNTs, the fabrication processes of CNT-based devices, etc. It will be of interest to physicists, chemists, material scientists, and engineers working on CNTs and related topics from both academic and industrial backgrounds. Finally, it should be mentioned that although we tried to avoid them, some errors would still remain in the book. We apologize in advance for those possible errors.

Acknowledgements

We are grateful to numerous colleagues who have helped us and provided precious suggestions for the composition of this monograph. There are too many to thank individually. We especially want to thank Professor Dong Xu of Shanghai Jiao Tong University, who proofread the book chapters and gave valuable pieces of advice for this book. We also thank Dr. Wei Zhang and Dr. Lin Hu, who helped to edit and collate the book manuscript. We also thank the Springer editor of this book, Dr. Claus Ascheron, for his support and patience throughout the preparation of the manuscript.

We acknowledge the support from The National Natural Science Foundation of China No. 60807008, Shanghai-Applied Materials Research and Development Fund No. 08520741500, Specialized Research Fund for the Doctoral Program of Higher Education (SRFDP) No. 200802481028, National Natural Science Foundation of China No. 50730008, and National Basic Research Program of China No. 2006CB300406 for the research work in the book.

Shanghai
February 2008

Changxin Chen
Yafei Zhang

Contents

Chapter 1
Carbon Nanotube Structure, Electronic, and Transport Properties

1.1 Introduction

The last few years witnessed a dramatic increase in nanomaterial and nanotechnology research. Among others, one of the most exciting fields to emerge is nanoelectronics [1–4], where a myriad of possibilities are appearing in the form of sensors [5], actuators [6], and transistors [7–9], each characterized by feature sizes of the order of a few nanometers. All this innovation has been fueled by the discovery of new materials and the invention of manufacturing methods that allow design and development at such a minute scale. Carbon nanotubes (CNTs) are at the forefront of these new materials, due to the unique mechanical and electrical properties that give them, e.g., exceptional strength [10] and conductivity [11]. Moreover, these nanotubes exhibit a tremendous current-carrying ability [12], potentially allowing for increased miniaturization of high-speed and high-power circuits.

CNT molecules have a cylindrical structure and are formed by one or more concentric, crystalline layers of carbon atoms. These atoms are assembled in hexagonal-lattice graphene sheets, which are rolled up into seamless tubes and named according to the number of concentric sheets as being either multi or single-walled nanotubes. Both kinds were originally observed experimentally, via transmission electron microscopy, decades ago, the research work on which in the field has subsequently increased dramatically.

CNTs exhibit excellent physical, mechanical, and chemical properties and so on, while this chapter mostly focuses on its electronic and transport characteristics. It is important to note that CNTs are being studied in a variety of fields to make use of their these properties. Nanotubes appear to be paving the way for myriad possibilities in the growing nanotechnology and emerging biotechnology industries, particularly in nanoelectronics in the form of sensors, transistors, and interconnections. One exciting possibility is the creation of nanometer-scale transistors, perhaps to be embedded, in the future, inside complex and miniscale electronic circuits that will make today's chips seem enormous in comparison. The potential application for CNTs in nano-optoelectronics also gives high expectation.

1.2 Structure of Carbon Nanotubes

1.2.1 Vector Notation and Structure Type of Carbon Nanotubes

The single-walled carbon nanotube (SWCNT) [13] is a hollow cylinder-shaped molecule with a diameter in the order of 1 nm. The best way to imagine the arrangement of the carbon atoms is to think of the SWCNT as a rolled up graphene sheet as sketched in Fig. 1.1. A graphene sheet is honeycomb lattice with a carbon atom in each vertex (Fig. 1.1a). This is also known as a trigonal-planar σ-bond framework, with a nearest-neighbor carbon–carbon distance a_{cc}, of approximately 0.144 nm along the bonds that are separated by 120°. The adjacent two atoms on the sheet constitute a cloud of delocalized π-oribitals surrounding the carbon cores – proper derivation of probability clouds would show that in reality the electrons form thicker, donut-shaped rings above the hexagonal lattice and that nearest donuts merge, thereby allowing electrons to move about the entire two-dimensional plane.

SWCNT are characterized by a wrapping (or chiral) vector $C_{\mathrm{h}} = na_1 + ma_2$, where a_1 and a_2 are the lattice basis vectors separated by 60° and the indices (n, m) are positive integers with $n \geq m \geq 0$. Figure 1.1b shows the chiral vector for a (4,2) nanotube, where the SWCNT is made by joining the ends of the chiral vector. Three categories of SWCNT are now defined: The armchair (n, n), the zigzag $(n, 0)$, and the chiral (n, m) with $n > m > 0$. The former two are named after the armchair- or zigzag-like pattern of the atoms around the tube circumference (see Figs. 1.1c and

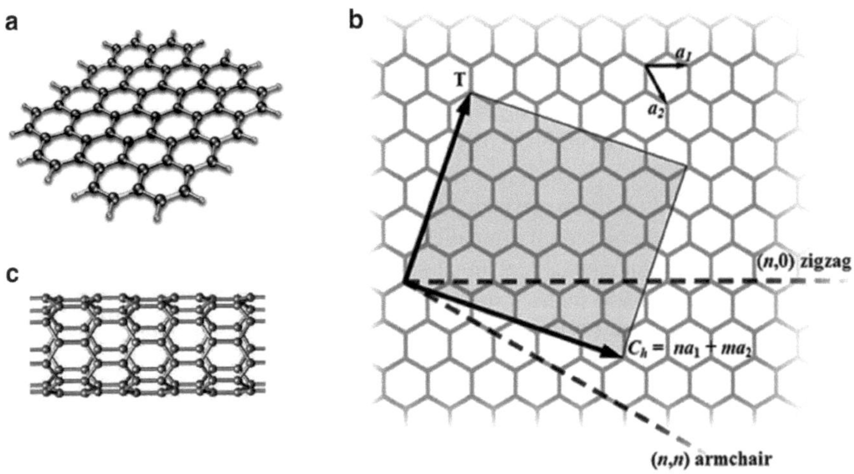

Fig. 1.1 (**a**) A layer of graphite (graphene), where the carbon atoms are arranged in a honeycomb lattice. (**b**) The example of folding a (4,2) SWCNT from the graphene sheet. C_{h} and T are the chiral and the translation vectors of the SWCNT, respectively, spanning the nanotube real space unit cell (*colored rectangle*). This unit cell differs from the graphene lattice unit cell spanned by the unit vectors a_1 and a_2. (**c**) A zigzag SWCNT. Note that it is different from the one sketched in (**b**)

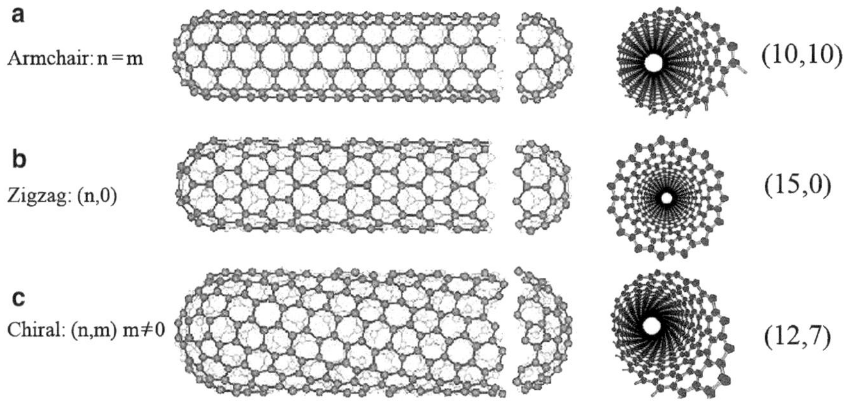

Fig. 1.2 The structure of different kinds of SWCNTs. (**a**) A (10,10) armchair SWCNT. (**b**) A (15,0) zigzag SWCNT. (**c**) A (12,7) chiral SWCNT

1.2a, b). In case of a chiral nanotube, the atomic arrangement seems to be spiral looking along the axis direction as shown in Fig. 1.2c. The smallest possible radius, a limit imposed by the requirement that the energy of the system in nanotube form be lower than that of the unrolled graphene equivalent, is thought to be 0.2 nm [14], whereas the upper limit on radius is in the several nanometer range. Quite remarkably, the electronic properties crucially depend on the exact folding of the graphene sheet. For completeness, it should be mentioned that the CNTs not only exist as SWCNT but also come in other species as ropes of SWCNT [15] or multiwalled carbon nanotube (MWCNT). The rope has many SWCNTs ($\sim$100) arranged in a triangular lattice, while the MWCNT consists of concentric shells of SWCNTs [16].

1.2.2 Metallic and Semiconducting Carbon Nanotubes

SWCNTs can be either metallic or semiconducting depending on the chiral vector as depicted previously in Fig. 1.3. Studies have shown that for given chiral vector directions, the electrical properties of the SWCNT can be predicted. If the difference of the vector integers, given by $n - m$, is a multiple of three, then the nanotube is metallic. Otherwise the electrical behavior of the tube will be semiconducting. The subsequent electrical behavior resulting from these integer combinations is shown in Fig. 1.3.

As shown in Fig. 1.3, the chiral vector integer pairs that produce metallic SWCNTs are represented as hollow circles, while semiconducting SWCNT producing pairs are shown as dark circles. MWCNTs tend to be metallic due to the low likelihood of all semiconducting nested nanotubes.

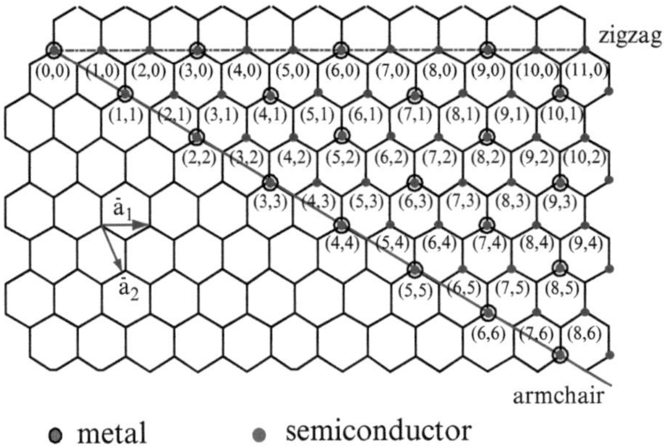

Fig. 1.3 Conductivity roadmap for SWCNTs indicating the conductivity of the nanotube produced with a chiral vector of given (n, m) integers. Image adapted from [16]

1.3 Electronic Properties of Carbon Nanotubes

1.3.1 Dispersion Relation of SWCNTs

The band structure of carbon nanotube can be derived from the band structures of grapheme. The two-dimensional dispersion relation of graphene can be evaluated using tight-binding approximation [17]:

$$E_{\text{graphene}}(k_x, k_y) = \pm t \left[1+4\cos\left(\frac{3k_x a_{cc}}{2}\right)\cos\left(\frac{\sqrt{3}k_y a_{cc}}{2}\right) + 4\cos^2\left(\frac{\sqrt{3}k_y a_{cc}}{2}\right)\right]^{1/2}, \tag{1.1}$$

where the positive and negative terms correspond to the symmetrical bonding and antibonding energy bands respectively, the k is wavevector, and t is the transfer integral (or nearest-neighbor parameter), the value of which is typically taken to have magnitude 2.8 eV. A plot of (1.1), representing a surface of allowed energies for the two-dimensional wavevector, is illustrated in Fig. 1.4, where the high-symmetry points are indicated by capital letters (K, M, and Γ). The K-points are degenerate near the Fermi energy ($E = 0$ in the plot) and indicate the zero-gap characteristic of conducting graphene. Near these K-points, as $k \to 0$, the dispersion relation is approximately cone-shaped and from the constant slope the Fermi velocity is given by [17]

$$\upsilon_F = \frac{3}{2\hbar} a_{cc} \left| t \right| .$$

The expression for the nanotube dispersion relation can be obtained by taking slices of the surface above, each cut being determined by the circumferential quantization. Again following the lattice vector definitions [18], we switch the wavevector notation $(k_x, k_y) \to (k_z, p)$, where the subscript z denotes the direction of transport,

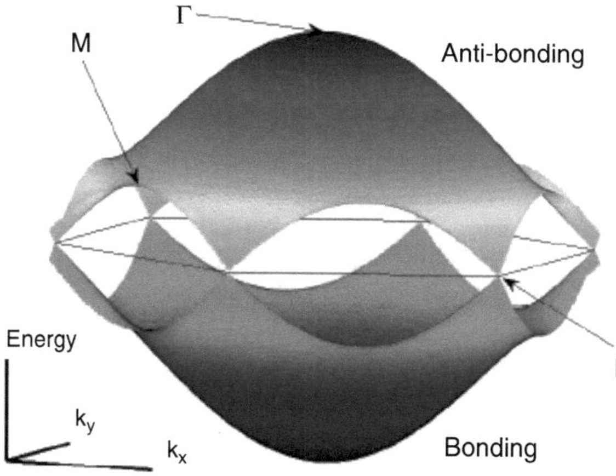

Fig. 1.4 Energy dispersion relation for graphene, from nearest-neighbor tight-binding calculation, using Slater–Koster approximation. Γ, M, and K are high-symmetry points, where the K-points lie on the plane of $E = 0$ (the Fermi level). Image adapted from [19]

and p is an integer representing the quantized modes obtained by imposing a periodic boundary condition in the circumferential (perpendicular to transport) direction. The nanotube dispersion relation is then given by

$$E_t\,(k_z, p) = \pm t(1 + 4\cos\gamma_1 \cos\gamma_2 + 4\cos^2\gamma_2)^{1/2}, \tag{1.2}$$

where the cosine arguments are given by

$$\gamma_1 = \frac{3a_{\rm cc}}{4}\frac{m-n}{\eta}k_z + \frac{\pi}{2}\frac{n+3m}{\eta^2}p,$$
$$\gamma_2 = \frac{3a_{\rm cc}}{4}\frac{m+n}{\eta}k_z + \frac{\pi}{2}\frac{3n-m}{\eta^2}p,$$

in which $\eta^2 = n^2 + m^2 + nm$, $-\pi < (3a_{\rm cc}\eta/d_{\rm g})k_z < +\pi$, $p = 0, \ldots, (2\eta/d_{\rm g} - 1)$, and $d_{\rm g} = \gcd(2n - m, 2m - n)$ [19].

1.3.2 Band Structure of SWCNTs

The band structure of the SWCNT is found by imposing periodic boundary conditions around the circumference of the tube, i.e., the wave function has to be single valued (only the plane wave part of the Bloch wave function written here) $e^{ik\cdot r} = e^{ik\cdot(r+C_{\rm h})}$, where k is a wavevector and r is a real space lattice vector of the graphene lattice. This leads to periodic boundary condition in momentum space

$$k \cdot C_{\rm h} = 2\pi p \tag{1.3}$$

where p is an integer. In other words, the k-vector projected onto the chiral vector $k_{\perp}$ (along the circumference) becomes quantized, while the k-vector $k_{||}$ along the tube axis is continuous for an infinite nanotube. The 1-D dispersion or band structure of a SWCNT is thus made of the energy bands related to different quantized values p as a function of $k_{||}$. Whether or not these quantization lines cross a K-point (Fig. 1.5) makes the SWCNT a metal or a semiconductor.

Based on the electronic properties of SWCNTs, three groups of SWCNTs can be defined as follows. The first group contains the armchair SWCNTs. These nanotubes

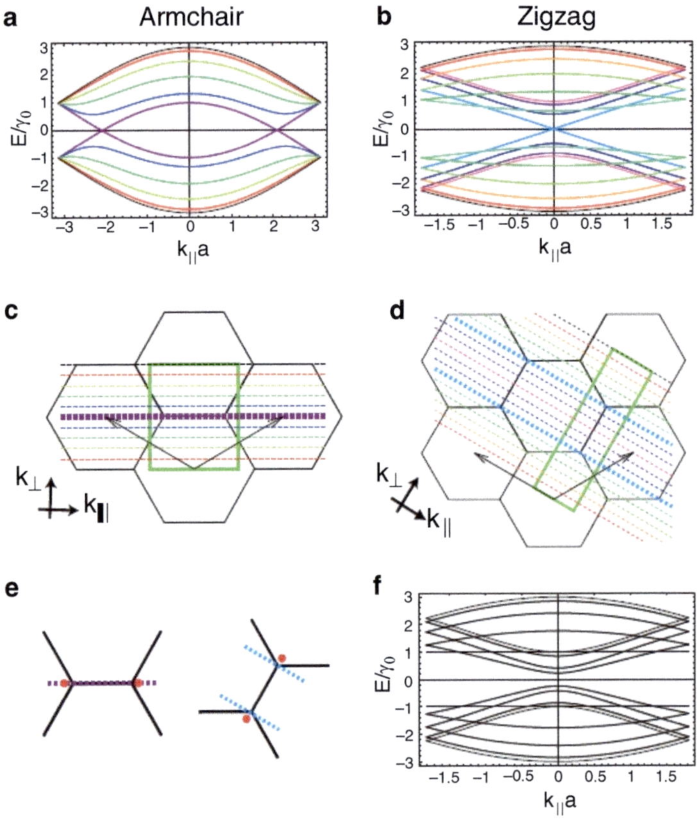

Fig. 1.5 (**a**) The band structure of a (5,5) armchair nanotube. The purple bands crossing the Fermi energy stem from the quantization line in (**c**), which crosses the K-points and makes this SWCNT metallic. (**b**) The band structure of a (9,0) zigzag nanotube. It has a doubly degenerate band crossing the Fermi level stemming from the two blue thick dashed lines in (**d**). (**c**) The dashed quantization lines in reciprocal space of a (5,5) SWCNT showing that one line crosses the K-points. The colors are related to the bands in (**a**). (**d**) Similar figure for a (9,0) zigzag tube. (**e**) When the curvature effects are taken into account the K-points move from the corners of the hexagon to the red dots. Armchair nanotubes thus stay metallic (*left*), while metallic zigzag nanotubes become small band gap semiconductors (*right*). (**f**) Some tubes are semiconducting with a bigger band gap as the (8,0) zigzag tube. Images adapted from [16]

are truly metallic and have two bands crossing at the Fermi level. Figure 1.5a shows the band structure for a (5,5) armchair nanotube. First note that two bands are crossing the Fermi level (or the valence and conduction band touch at the Fermi level) making it a metal. The bands stem from the quantization lines drawn in Fig. 1.5c in the reciprocal lattice. The corners of the hexagons are the K-points, where the conduction and the valence band of the graphene dispersion touch. One of the quantization lines (thick purple dashed line) passes through two K-points making the tube metallic. Note that the degeneracy of all bands is double, except the top conduction band, the bottom valence band (black dashed quantization line), and the two bands crossing the Fermi energy (purple quantization line). Figure 1.5b shows the band structure for a (9,0) zigzag, tube which is metallic judging from the (degenerate) band crossing the Fermi level. The bands stem from the quantization lines drawn in Fig. 1.5d. It is seen that the bands touching at the Fermi level are two times degenerate (blue thick dashed lines) [16].

However, the aforementioned band structure is calculated from the dispersion of a sheet of graphene, while the nanotube has a curvature around the circumference of the tube. The curvature slightly modifies the band structure by moving the K-points [20].

In the armchair case, the SWCNTs are still metallic because the shift of the K-points happens along the quantization lines. Figure 1.5e (left) shows the new positions of the K-points by red dots. This is not the case for the metallic zigzag nanotube. The shift of the K-points is not along quantization lines and thus turns the metallic zigzag nanotube into a small band gap semiconducting nanotube. Note that the K-points are located at different sides of their respective quantization line in Fig. 1.5e (right). The two lowest lying subbands are still degenerate, but the slope of the dispersion on the two quantization lines with respect to the perpendicular k-vector $k_\perp$ has the opposite sign. Similar transitions to a small band gap semiconductor happens for other metallic zigzag tubes and chiral metallic tubes for which $n - m \approx 0(\text{mod}3)$. The effect is the stronger for small diameter tubes due to their bigger curvature. The gap will be in the order of tens of millielectron volts.

The last category of tubes for which $n - m \neq 0(\text{mod}3)$ is the semiconducting as seen in Fig. 1.5f, which shows a (8,0) zigzag SWCNT. In the reciprocal space, the quantization lines do not cross the K-points (not shown). It has a band gap in the order of $\sim$1 eV and no available state at the Fermi level. Curvature effect also modifies the band gap obtained by quantization of the graphene sheet, but at a much smaller energy scale.

Therefore, the SWCNT will either be metallic, a small band gap semiconductor, or a large band gap semiconductor. The armchair SWCNTs are a group of truly metallic conductors with two bands crossing the Fermi level. The initially metallic zigzag and chiral SWCNTs ($n - m \approx 0(\text{mod}3)$) without curvature effects turn into small band gap semiconductors, when curvature effects are included. Typical band gaps are in the order of tens of millielectron volts. Finally, a group of zigzag and chiral SWCNTs is semiconducting ($n - m \neq 0\,(\text{mod}3)$) with bigger band gaps. The band gap of these tubes are in the order of $\sim$1 eV and scales as $E_{\text{gap}} \sim 1/d$, where d is the diameter of the SWCNT [21].

1.3.3 Density of State for SWCNTs

Due to the 1-D nature of their band structure, SWCNTs exhibit peaks or spikes in their density of state (DOS) diagrams as shown in Fig. 1.6.

The DOS diagrams shown in Fig. 1.6a, b are representative of semiconducting and metallic CNTs respectively. The peaks in the DOS diagrams are called van Hove singularities and are representative of the energy levels where the electrons tend to exist in the orbitals. One distinguishing difference between the semiconducting and the metallic DOS diagrams is that electrons can exist at zero energy. In the outer singularities of the semiconducting tube, there exists a need to transition between peaks, as opposed to the metallic tube where electrons can move more readily. The band gaps between the van Hove singularities seen in Fig. 1.6 can be estimated using the theoretical prediction,

$$E = \frac{k a_{\rm cc} \gamma_0}{d}. \tag{1.4}$$

In this relation, E is the transition energy, a_{cc} is the nearest-neighbor carbon–carbon distance, 0.144 nm, γ_0 is the nearest-neighbor carbon–carbon interaction energy, 2.9 eV [22], d is the SWCNT diameter, and k is an integer constant. The values of k can be 2, 4, and 8 for the first, second, and third van Hove singularities in semiconducting tubes and 6 and 12 for the first and second van Hove transitions in metallic tubes respectively. The ability to create SWCNTs in large quantities with a common chirality and electrical properties is still the largest problem faced in the CNT synthesis field [23].

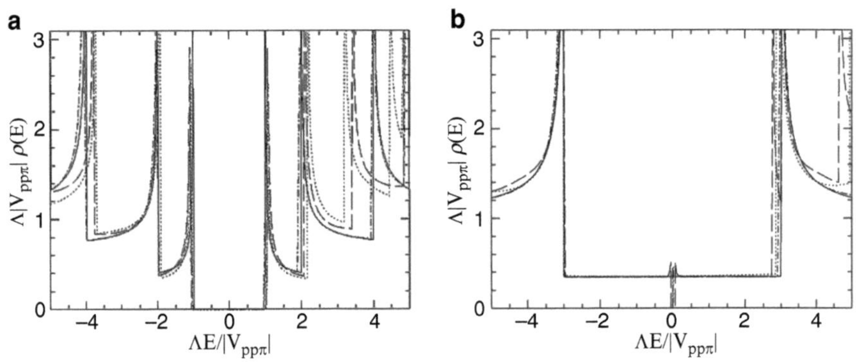

Fig. 1.6 Density of states diagrams for a semiconducting CNT (**a**) and a metallic CNT (**b**) taken from [21]. The spikes in the figure represent energy levels at which electrons tend to exist and are called van Hove singularities. The metallic tube (**b**) can readily move electrons between the two lowest energy levels. The energy band gap seen in (**a**) between the lowest level states is indicative of a semiconducting tube (**a**)

1.4 Transport in Single-Walled Carbon Nanotubes

According to Landauer's formula [24] that is very useful in describing mesoscopic transport, the conductance is

$$G = G_0 NT, \tag{1.5}$$

where the factor T represents the average probability that an electron injected at one end of the conductor will transmit to the other end, which is determined by the properties of the conductor and also the contacts. $G_0 = 2e^2/h = (12.9\,\mathrm{k\Omega})^{-1} = 75\,\mu\mathrm{S}$ is the conductance quantum including spin degeneracy. N is the number of available conducting subbands inside the conductor.

According to the (1.5), in the ideal case, a ballistic conductor with perfect contacts (i.e. transparency = 1, no back reflection), the transmission probability (T) is unity, the conductance of the system with N conducting subbands is NG_0 due to the contact resistance.

In the nonideal case, the transmission is reduced due to back-scattering in the narrow conductor and imperfect contacts. Assume that the transmission coefficients of all the conducting channels are T $(0 \le T \le 1)$, then $G = G_0 NT$. This resistance can be divided into two parts: the contact resistance originating in the transition to the contacts and the residual scatter's resistance due to the narrow conductor and imperfect contacts:

$$G^{-1} = (NTG_0)^{-1} = G_\mathrm{c}^{-1} + G_\mathrm{s}^{-1} \tag{1.6}$$

here $G_\mathrm{c}^{-1} = 1/NG_0$ and $G_\mathrm{s}^{-1} = 1/\,(NG_0)\cdot(1-T)\,/\,T$ [25]. The contact resistance, G_c^{-1}, is inversely proportional to the number of available conducting subbands N. The contact resistance of a single-band conductor is about 12.9 KΩ. And if N is very large, the contact resistance may become very small. In that case, the scatter resistance G_s^{-1} dominates, which is determined by the properties of the scatter, that is, the transmission probability, T. In the case of a ballistic conductor with imperfect conducts, the reduced transmission probability T is only due to back-scattering in the contacts. Similarly, in the case of a conductor with perfect contacts, the reduced transmission probability, T, is due to back-scattering in the conductor only.

For the armchair nanotube, the valence and the conduction band cross at the Fermi level. The tube has two bands that can carry currents close to E_F, or two conducting channels. In each band, we can put two electrons with opposite spin. Thus, an ideal armchair nanotube with perfect contacts to a metal has a conductance of $2G_0 = 4e^2/h$ [26].

Figure 1.7a, b show the conductance (full line) and the band structure of a metallic (5,5) armchair nanotube calculated from the first principles [27]. If the Fermi level is shifted, e.g., to $\Delta E_F = -0.2\,\mathrm{eV}$ (upper *dashed line* in (b)) only the two crossing bands with $m = n = 5$ conduct the current as discussed earlier. At even larger $\Delta E_\mathrm{F} = -1.8\,\mathrm{eV}$ (lower dashed line), we find four k points in the valence bands intersecting the Fermi level. Since the lower-lying valence bands are twofold degenerate, the conductance is now $6G_0$. The conductance is reduced by G_0 at $\Delta E_\mathrm{F} = -2\,\mathrm{eV}$, because one of the $m = 5$ bands then is empty as can be seen in Fig. 1.7b. The dashed lines in Fig. 1.7a are for a quasimetallic (9,0) zigzag

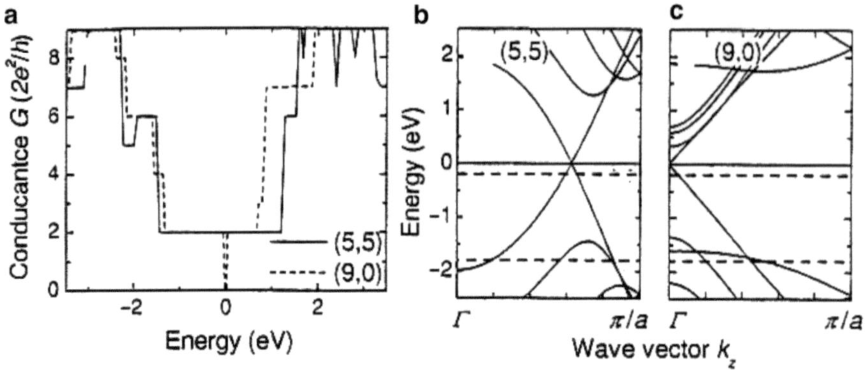

Fig. 1.7 (**a**) Conductance of a (5,5) armchair (*full line*) and a (9,0) zigzag nanotube (*dashed line*). The conductance is quantized in multiples of G_0. While the truly metallic armchair nanotube has a finite conductance at the Fermi level, the conductance of the (9,0) zigzag tube drops to zero at the Fermi level. The energy scale is such that the Fermi level is at 0. (**b, c**) Band structure of (**b**) a (5,5) and (**c**) a (9,0) nanotube. The *dashed lines* represent the Fermi level shifted to $\Delta E_F = -0.2$ and -1.8 eV. Images adapted from [26]

nanotubes; its band structure is shown in Fig.1.7c. Note the secondary gap at E_F, which comes from the curvature of the nanotube wall. At very low temperatures this gap prevents transport in nonarmchair $(n-m)/3$ integer nanotubes. In a zigzag tube all bands are two-folded degenerate close to E_F and the critical point is the Γ point. At $\Delta E_F = -0.2$ eV, one degenerate band gives rise to a conductance of $2G_0$; at -1.8 eV, we find three bands, i.e., $G = 6G_0$. Finally, semiconducting SWCNTs have band gaps between $\approx$0.5 and 1 eV; except for the large band gap, they behave exactly as the quasi-metallic (9,0) zigzag tube in Fig. 1.7a, c.

For the metallic and semiconducting SWCNTs, two different types of behaviors of the room-temperature conductance (zero bias) versus gate voltage are exhibited. In Fig. 1.8a, the conductance is independent of the gate voltage, which is attributed to metallic nanotubes. The crossing of the valence and conduction band in the metallic nanotubes provides conducting electrons independently of the gate voltage. This is schematically shown by the inset in Fig. 1.8a. The conductance in semiconducting nanotubes, in contrast, changes by orders of magnitude when changing V_g (see Fig. 1.8b). In the semiconducting nanotubes, an asymmetry between the turn-on gate voltage below and above the Fermi level was observed. The tubes became conducting for negative V_g, at a lower absolute value than for positive V_g, which was naturally attributed to an intrinsic p-type doping of the tube [28, 29]. However, a more careful study by Heinze et al. showed that the asymmetry between the two conducting channels is not due to a doping of the tube by O_2, but rather due to Schottky barriers that form at tube–metal interface [30]. This can be illustrated from the energy band diagram of the nanotube–metal contact. Figure 1.9 shows the band structure of a nanotube contacted to a metal electrode (full lines), the work function of which was assumed to be larger than that of the tube. Far away from the junction the Fermi level of the nanotube is in the middle between the valence

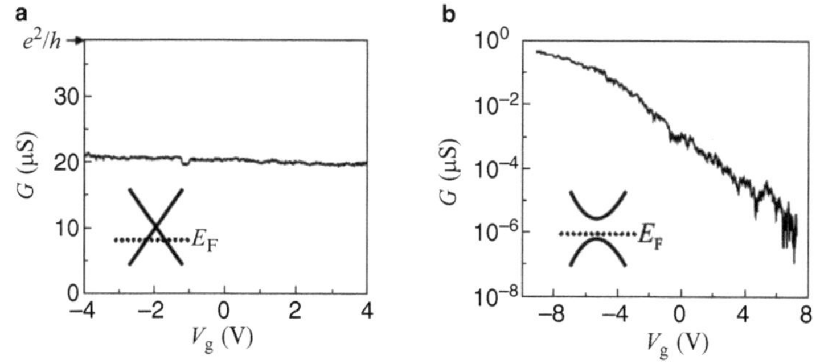

Fig. 1.8 Conductance of (**a**) a metallic and (**b**) a semiconducting nanotube at room temperature. *Insets*: schematic drawing of the expected dispersion $E(k)$ near the Fermi level E_F for a metallic tube (**a**) and a semiconducting tube (**b**). Images adapted from [31]

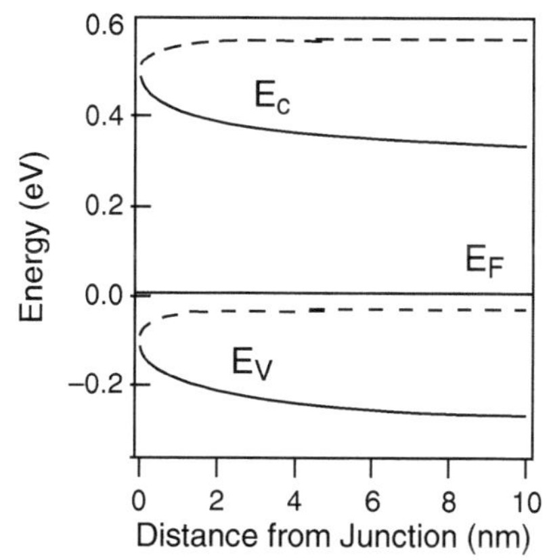

Fig. 1.9 The SWCNT band bending in the case that the work function of the metal electrode is +0.2 eV larger than the energy level of the SWCNT median band (the Fermi level of the SWCNT). The *full lines* correspond to zero gate voltage; the *dashed lines* are for −500 mV gate voltage. Images adapted from [30]

and conduction band. Applying a positive or negative gate voltage, we can switch the nanotube into the conducting state. At negative voltage (dash lines) the Schottky barrier is thinner than for a positive gate voltage. Consequently, a larger current flows through the barrier for the p channel than for the n channel at the same absolute V_g. Owing to the Schottky barriers at the tube–metal interface, the operation of nanotubes field-effect transistors is not based on modulating the charge density in the channel, but rather on a modulation of the contact barriers [26, 30].

1.5 Conclusion

In this chapter, we have had a look at the structure of CNTs at first. SWCNTs can be categorized into three types of armchair tube (n,n), zigzag tube $(n,0)$, and chiral tube (n, m) according to their chiral vector. They can be either metallic or semiconducting depending on the chiral vector. If the difference of the vector integers, given by $n - m$, is a multiple of three, then the nanotube is metallic. Otherwise the electrical behavior of the tube will be semiconducting.

The electronic properties including the dispersion relation, band structure, and density of state for SWCNTs have also been introduced. Based on the band structure, SWCNTs can be metallic, a small band gap semiconductor, or a large band gap semiconductor. The armchair SWCNTs are a group of truly metallic conductors with two bands crossing the Fermi level. The initially metallic zigzag and chiral SWCNTs ($n - m \approx 0(\mathrm{mod}3)$) without curvature effects turn into small band gap semiconductors, when curvature effects are included. Another group of zigzag and chiral SWCNTs is semiconducting ($n-m \neq 0(\bmod 3)$) with bigger band gaps. Due to the 1-D nature of their band structure, SWCNTs exhibit peaks or spikes in their density of state (DOS) diagram.

The transport properties of metallic and semiconducting SWCNTs, including the conductance and its dependency on the gate voltage, are also discussed. It is shown that the conductance in the metallic nanotube is independent of the gate voltage, and in contrast, the conductance of the semiconducting nanotubes changes by orders of magnitude when changing gate voltage. The asymmetry between the conductance of electrons and holes for the semiconducting nanotubes can be attributed to the different height of Schottky barriers for electrons and holes formed at tube–metal contacts.

References

1. C.R. Martin, Science **266**, 1961 (1994)
2. R.F. Service, Science **294**, 2442 (2001)
3. M. Turner, V.B. Golovko, O.P.H. Vaughan, P. Abdulkin, A.B. Murcia, M.S. Tikhov, B.F.G. Johnson, R.M. Lambert, Nature **454**, 981–983 (2008)
4. M. Law, J. Goldberger, P. Yang, Ann. Rev. Mater. Res. **34**, 83 (2004)
5. Y.S. Zhang, K. Yu, D.S. Jiang, Z.Q. Zhu, H.R. Geng, L.Q. Luo, Appl. Surf. Sci. **242**(1–2), 212 (2005)
6. R.H. Baughman, C.X. Cui, A.A. Zakhidov, Z. Iqbal, J.N. Barisci, G.M. Spinks, G.G. Wallace, A. Mazzoldi, D.D. Rossi, A.G. Rinzler, O. Jaschinski, S. Roth, M. Kertesz, Science **284**(5418) 1340 (1999)
7. A. Bachtold, P. Hadley, T. Nakanishi, C. Dekker, Science **294**(5545) 1317 (2001)
8. H.W.Ch. Postma, T. Teepen, Z. Yao, M. Grifoni, C. Dekker, Science **293**(5527), 76 (2001)
9. A. Javey, J. Guo, Q. Wang, M. Lundstrom and H.J. Dai, Nature **424**, 654 (2003)
10. E.T. Thostenson, Z.F. Ren, T.W. Chou, Comp. Sci. Technol. **61**, 1899 (2001)
11. S. Berber, Y.K. Kwon, D. Tomanek, Phys. Rev. Lett. **84**, 4613 (2000)
12. S.Y. Moon, W. Song, N. Kim, J.S. Lee, P.S. Na, S.G. Lee, J.W. Park, M.H. Jung, H.W. Lee, K. Kang, C.J. Lee, J. Kim, Nanotechnology **18**, 234201 (2007)
13. S. Iijima, T. Ichihashi, Nature **363**, 603 (1993)
14. X. Blase, X. Lorin, E. Benedict, L. Shirley, S.G. Louie, Phys. Rev. Lett. **72**, 1878 (1994)
15. A. Thess, R. Lee, P. Nikolaev, H. Dai, P. Petit, Science **273**, 483 (1996)
16. K. Grove-Rasmussen, Ph.D. thesis (University of Copenhagen, 2006)
17. A. Godoy, Z. Yang, U. Ravaioli, F. Gamiz, J. Appl. Phys. **98**, 013702 (2005)
18. R. Saito, G. Dresselhaus, M.S. Dresselhaus, *Physical Properties of Carbon Nanotubes* (Imperial College Press, London, 1998)
19. C. Leonardo de Camargo, Ph.D. Thesis (The University of British Columbia, 2006)

20. L. Balents, P.A. Fisher, Phys. Rev. B. **55**, R11973 (1997)
21. J.W. Mintmire, C.T. White, Phys. Rev. Lett. **81**, 2506 (1998)
22. S.D.M. Brown, P. Corio, A. Marucci, M.S. Dresselhaus, M.A. Pimenta, K. Kneipp, Phys. Rev. B **61**, R5137 (2000)
23. E.O. Sunder, M.Sc. thesis (Georgia Institute of Technology, 2006)
24. R.J. Landauer, Phys. Condens. Matter. **1**, 8099 (1989)
25. S. Datta, *Electronic Transport Properties in Mesoscopic Systems* (Cambridge University Press, Cambridge, 1995)
26. S. Reich, C. Thomsen, J. Maultzsch, *Carbon Nanotubes: Basic Concepts and Physical Properties* (Wiley, New York, 2004), p 85
27. M. Brandbyge, J.L. Mozos, P. Ordejon, J. Taylor, K. Stokbro, Phys. Rev. B **65**, 165401 (2002)
28. S.J. Tans, M.H. Devoret, H. Dai, A. Thess, R.E. Smalley, C. Dekker, Nature **386**, 474 (1997)
29. R. Martel, T. Schmidt, H.R. Shea, T. Hertel, P. Avouris, Appl. Phys. Lett. **73**, 2447 (1998)
30. S. Heize, J. Tersoff, R. Martel, V. Derycke, J. Appenzeller, P. Avouris, Phys. Rev. Lett. **89**, 106801 (2002)
31. J. Nygard, D. Cobden, M. Bockrath, P. McEuen, P. Lindelof, Appl. Phys. A: Mater. Sci. Process. **69**, 297 (1999)

Chapter 2
Synthesis and Purification of Carbon Nanotubes

2.1 Introduction

Over the last one and a half decades, a great amount of research on the synthesis of carbon nanotubes has been carried out [1–5] since the first pioneering report of the discovery of multiwalled carbon nanotubes (MWCNTs) by Iijima [6]. Carbon nanotubes (CNTs) are generally produced by three main techniques: arc discharge, chemical vapor deposition (CVD), and laser ablation, although scientists are researching more economic ways to produce these structures. In arc discharge, a vapor is created by an arc discharge between two carbon electrodes with or without catalyst. Nanotubes self-assemble from the resulting carbon vapor. CVD synthesis is achieved by putting a carbon source in the gas phase and using an energy source, such as plasma or a resistively heated coil, to transfer energy to a reactive gaseous carbon molecule. This reactive carbon will deposit and grow on the catalyst (usually a first row transition metal such as Ni, Fe, or Co). The CVD method can result in MWCNTs or single-walled carbon nanotubes (SWCNTs). The CVD method is easy to scale up, which favors commercial production. For the laser ablation method, a mixture of carbon and transition metals such as Co and Ni is vaporized directly by a laser impinging on a metal–graphite composite target. In contrast to the arc discharge method, direct vaporization allows a far greater control over growth conditions, permitting continuous operation and preparation of nanotubes in higher quality. The laser ablation method tends to produce a small amount of clean nanotubes, whereas the arc discharge method generally produces larger quantities of nanotubes but also much more impurity.

The as-produced CNT soot contains a lot of impurities. The main impurities in the soot are graphite (wrapped up) sheets, amorphous carbon, metal catalyst, and the smaller fullerenes. These impurities will interfere with most of the desired properties of the SWCNTs. In the fundamental research as well as practical applications, it is preferred for one to obtain SWCNTs as pure as possible without changing their properties.

Aiming at removing different impurities, several methods have been attempted to purify CNTs. The metal catalyst in the raw product can be removed by their chemical reaction with the acid such as hydrochloric acid or nitric acid. The removal of amorphous carbon, graphite sheets, and small fullerenes etc. is usually based on one or more competitive oxidation reactions [7]. Many different oxidation routes have been used, including mixed mineral acid, hydrogen peroxide and potassium permanganate refluxes, gas phase oxidation [8, 15], and microwave oxidation [9]. Thereinto, the acid treatment [10, 11] is the common factor in the majority of purification regimes. The classic treatment for MWCNTs involves a (mixed) acid reflux followed by a water wash [10,11]. For SWCNTs, either a similar short mixed nitric/sulfuric reflux [12] or a longer (more dilute) nitric acid reflux are commonly used [10]. It is well known that these oxidations can damage the framework of the nanotubes, introducing chemical defects into the nanotube structure, both at the ends and at sites along the tube length [11]. In this chapter, a simple, effective, and minimally-destructive method for purifying SWCNT will be introduced and discussed.

2.2 Synthesis Methods of Carbon Nanotubes

2.2.1 Arc Discharge Method

Among the researches on CNTs, SWCNTs are particularly active and fruitful because of their especially interesting properties. Initially, about 1–2% yields of SWCNTs were found in 1993 in soot generated in an arc struck between graphite electrodes containing a few Fe [6], Co [13], Ni [14]. Since then, the preparation of high-quality SWCNTs on a large scale has been the objective of many research endeavors [15–19]. The arc-discharge method, which was initially used for preparing C_{60} fullerenes, is the most common and perhaps the easiest way to produce carbon nanotubes, for it is rather simple to undertake. The method creates nanotubes through arc-vaporization of two carbon rods placed end to end, separated by approximately 1 mm, in an enclosure that is usually filled with inert gas (helium, argon) at low pressure (between 50 and 700 mbar). The metal powders such as Fe, Co, Ni, Y, and their compound etc. are added into carbon rods as the growth catalysis. A direct current of 50–100 A driven by approximately 20 V creates a high-temperature discharge between the two electrodes. The discharge vaporizes one of the carbon rods and forms a small rod-shaped deposit on the other rod. Producing nanotubes in high yield depends on the uniformity of the plasma arc and the temperature of the deposit form on the carbon electrode [38]. Figure 2.1 is a schematic apparatus drawing of the arc-discharge used for producing carbon nanotubes.

Shi et al. reported [19] a DC arc discharge method producing a large amount of SWCNTs in the soot with Y–Ni alloy instead of metallic yttrium and nickel mixture. The SWCNTs produced by DC arc discharge method is shown in detail as follows. A ø3 × 70 mm hole was drilled in a ø6 × 300 mm spectrally pure graphite rod and filled with 1:1 molar ratio of graphite and Y–Ni alloy (YNi_2) powder. It is desirable to synthesize SWCNT using YNi_2 instead of their metal mixture, because Y–Ni alloy is easily ground into powder and stable in the atmosphere. The cathode

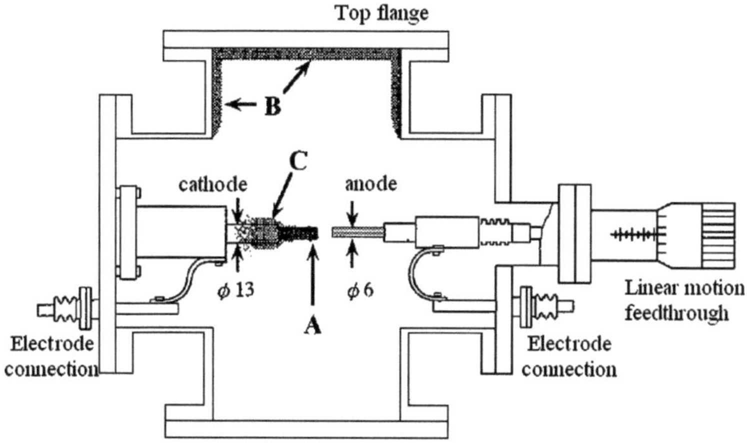

Fig. 2.1 Schematic apparatus drawing of the arc evaporator used for producing carbon nanotubes. A slag-like hard deposit (**A**) grown on the end of the cathode, soot (**B**) deposited on the walls of the evaporator, and soot (**C**) grown on surfaces of the cathode were separately collected. Images are adapted from [17]

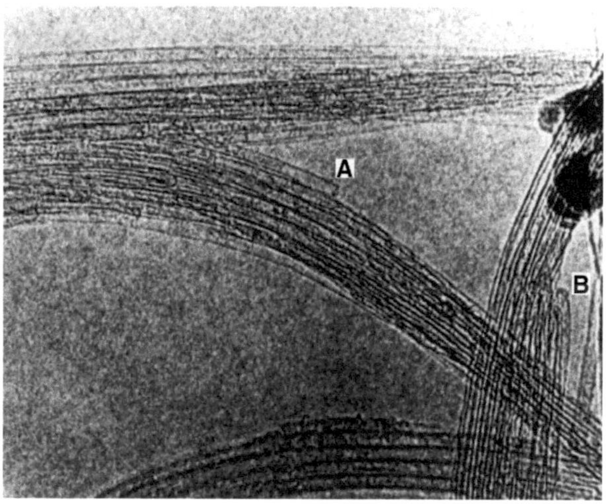

Fig. 2.2 HRTEM images of the sample obtained by arc discharge method, in which the random SWCNT bundles and SWCNTs with flat (**A**) and hemispherical cap (**B**). Image adapted from [19]

was a ø10 mm graphite rod with a sharp end toward the anode in order to decrease the formation of cathode deposit. The arc was generated by a current of 40–100 A in a helium atmosphere at a pressure of 100–700 torr. When the discharge was running, the distance between two electrodes was maintained constant at ~5 mm by continuously translating the anode. After arc discharge, the soot was extracted by CS_2, washed with 1:1 hydrochloric acid, and dried at 100°C to remove most of the fullerens and catalysts. Figure 2.2 is the high-resolution transmission electron microscopy (HRTEM) images of the product. From the image, it can be found that most SWCNTs exist in bundles, each bundle consisting of tens of SWCNTs [20]. The SWCNTs have nearly the same diameter of about 1.3 nm with a length of several microns. It can be seen that there are two types of caps at the end of SWCNT, one being flat in the central part of the image (a) and the other being hemispherical in the right part of the image (b) [21].

Liu et al. [5] reported a hydrogen arc discharge method as shown in Fig. 2.3, by which large-scale high-quality SWCNTs can be semicontinuously synthesized at low cost, since the employed H_2 rather than expensive He as the buffer gas, designed a novel reactor and introduced a sulfur-containing SWCNT-growth promoter. The synthesized carbon nantouebs in semicontinuous method is shown in Fig. 2.4a. It is difficult to measure the length of the bundles accurately, which is estimated to be several, or even up to tens of microns. After mounting on a holy copper grid, the as-prepared thin film substance was observed by TEM as shown in Fig. 2.4b.

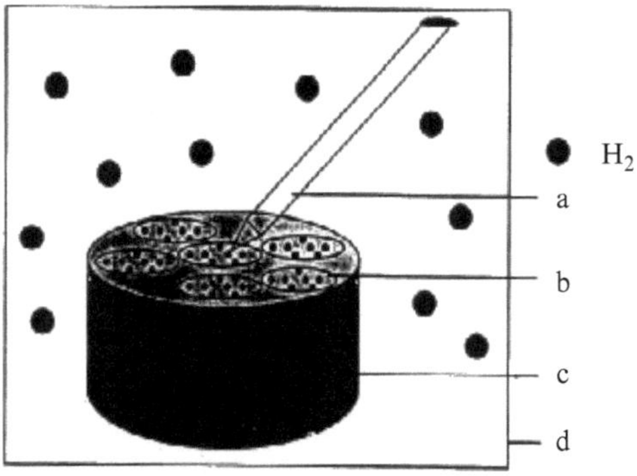

Fig. 2.3 Schematic diagram of the apparatus used for synthesizing SWCNTs by the semicontinuous hydrogen arc discharge method. (**a**) Graphite cathode; (**b**) mixture of graphite powder and catalysts; (**c**) graphite anode; (**d**) chamber wall. As we can see, the two electrodes make an oblique angle, and this point is different from the conventional apparatus, in which the cathode and the anode are vertically opposite [6, 22, 23]. Image adapted from [5]

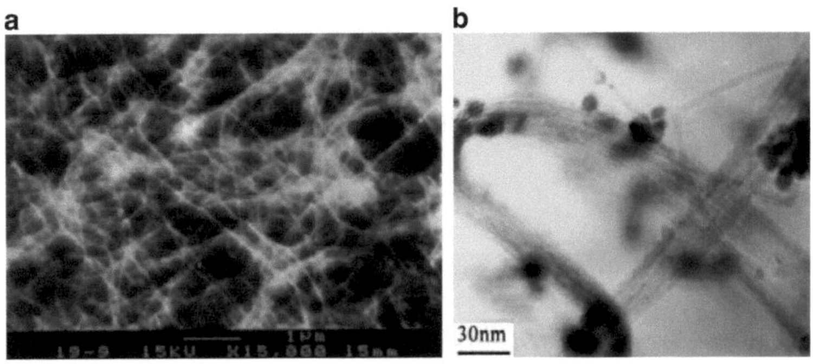

Fig. 2.4 (**a**) SEM image of the as-prepared web-like substance. The entangled SWCNT bundles are usually curved or circled (scale bar, 1 μm) (**b**) TEM image of the as-prepared thin film substance; two SWCNT bundles comprising tens of packed SWCNTs can be seen. Images adapted from [5]

2.2.2 Chemical Vapor Deposition (CVD) Method

Another main method for growing CNTs is the CVD method. Excellent alignment [24], as well as positional control on nanometer scale [25], can be achieved by using CVD method. Control over the diameter, as well as the growth rate of the nanotubes, can also be maintained. In the CVD method, the appropriate metal catalyst can preferentially grow single-walled rather than multiwalled nanotubes [26].

The CVD method differs from arc discharge techniques, which does not utilize a solid carbon source. The synthesis is essentially a two-step process consisting of a catalyst preparation step followed by the actual synthesis of the nanotube. The catalyst is generally prepared by sputtering a transition metal onto a substrate and then using either chemical etching or thermal annealing to induce catalyst particle nucleation. Thermal annealing results in cluster formation on the substrate, from which the nanotubes will grow. Ammonia may be used as the etchant [26–29]. The temperatures for the synthesis of nanotubes by CVD are generally within the 650–900°C range [24–26, 29]. Typical yields for CVD are approximately 30%. In the second step, instead the carbon is gaseous and sources include methane, carbon monoxide and acetylene, which are split apart by an external energy source, e.g., a plasma or a heated coil. The heated reactive carbon diffuses toward the substrate surface where catalyst particles, e.g., Ni, Fe, or Co have previously been placed in the first step. The carbon binds to the catalyst particle and depending on the growth parameters MWCNT or SWCNT may start growing. The advantage of CVD method is the possibilities for large up scaling, simple processing and control over length and diameter of tubes [21].

The growth of CNTs is performed in a CVD furnace (Fig. 2.5), which allows for the simultaneous injection of three gases argon (Ar), hydrogen (H_2), and methane (CH_4) from the gas inlet. The optimal growth conditions have proved to be at temperatures in the range 850–950°C, and with a mixed gas flow consisting of Ar: 1.1 L min^{-1}, H_2: 0.1 L min^{-1}, and CH_4: 0.5 L min^{-1}. Before the actual CVD growing process, there are two preliminary processes before growth can be initiated. First, the sample is placed in the center of the furnace and rinsed with argon (2.8 L min^{-1}) for 3 min, making sure that the growing environment is free of growth contaminating gasses. Then the furnace and sample is heated under constant gas

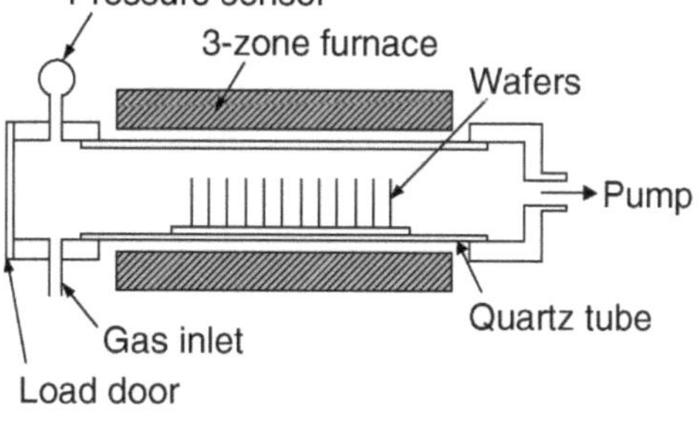

Fig. 2.5 Schematic diagram of the CVD for synthesis of carbon nanotubes

flow consisting of Ar: 1.1 L min^{-1} and H_2: 0.1 L min^{-1} to reach the desired growing temperature 850–950°C. At the desired temperature, the gas flow is changed by adding the carbon supply gas methane, giving a gas flow of Ar: 1.1 L min^{-1}, H_2: 0.1 L min^{-1}, and CH_4: 0.5 L min^{-1}. Growth is allowed for 10 min before turning off both hydrogen and methane gasses leaving only argon to flow during the cooling process. The cooling process is sped up by air cooling on the underside of the furnace chamber. After growth, the quality of the nanotubes is tested by inspection using SEM in the test areas. If the tubes in the test area are many and short, or there are not enough tubes, the growth cycle is repeated at a different temperature. High temperatures have shown to yield fewer but longer and thinner tubes, whereas lower temperatures give shorter and sometimes multiwalled tubes [21].

The type of catalyst support is found to be able to control the formation of individual or bundled SWCNTs. Previous researches [30] have shown that catalysts supported on crystalline alumina nanoparticles produce abundant individual SWCNTs and small bundles, while catalysts supported by amorphous silica particles produce only SWCNT bundles. Figure 2.6a, b show individual SWCNTs and bundles of SWCNTs synthesized with the Fe_2O_3/alumina catalyst and Fe_2O_3/silica catalyst respectively.

Fonseca et al. reported the synthesis of MWCNTs [27] and bundles of aligned and straight 1 nm diameter nanotubes over supported catalysts by the decomposition of hydrocarbons. The amount and dimension of the catalytic particles dispersed on the support, which was not easy to control precisely, was found to be important in controlling the shape of the nanotubes that are produced. Cheng et al. reported [28] a novel synthesis method for SWCNTs using an improved floating catalyst approach

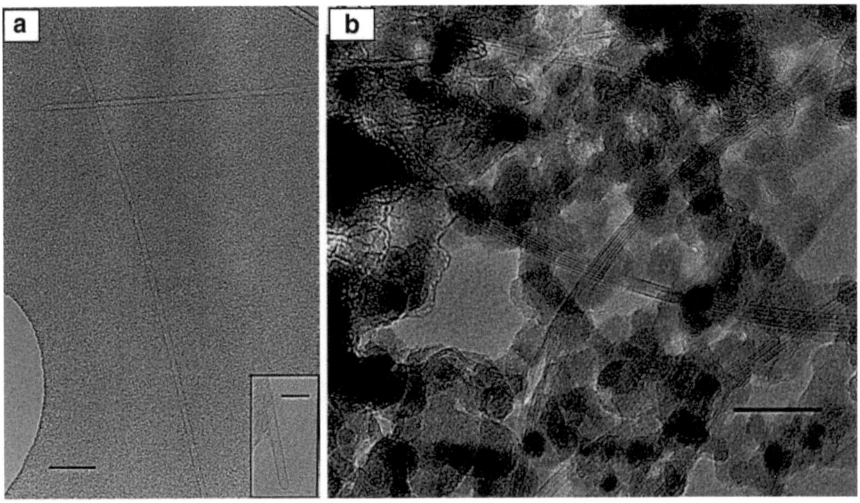

Fig. 2.6 (**a**) High-resolution TEM image of a $d = 5$ nm individual SWCNT produced on the Fe_2O_3/alumina catalyst. Scale bar: 50 nm. *Inset*: Shows the dome-closed end of a $d = 3$ nm SWCNT. Scale bar: 10 nm. (**b**) High-resolution TEM image of SWCNT bundles synthesized with the Fe_2O_3/silica catalyst. Scale bar: 50 nm. Images are adapted from [30]

in which benzene was catalytically pyrolyzed at 1,100–1,200°C. In contrast to the electric-arc and laser vaporization methods, this method allows lower growth temperaturek and easier control over the growth conditions, permits semicontinuous or continuous preparation, and produces SWCNTs at high purity, large quantity, and at relatively low cost.

2.2.3 Laser Ablation Method

In 1995, Smalley's group [31] at Rice University reported the synthesis of CNTs by laser vaporization. The laser vaporization apparatus used by Smalley's group is shown in Fig. 2.7 [32]. A pulsed [33,34], or continuous [35] laser is used to vaporize a graphite target in an oven at 1,200°C. The main difference between continuous and pulsed laser is that the pulsed laser demands a much higher light intensity (100 kW cm^{-2} as against 12 kW cm^{-2}). The oven is filled with helium or argon gas in order to keep the pressure at 500 torr. A very hot vapor plume forms, then expands and cools rapidly. As the vaporized species cool, small carbon molecules and atoms quickly condense to form larger clusters, possibly including fullerenes. The catalysts also begin to condense, but more slowly at first, and attach to carbon clusters and prevent their closing into cage structures [36]. Catalysts may even open cage structures when they attach to them. From these initial clusters, tubular molecules grow into SWCNTs until the catalyst particles become too large or until conditions have cooled sufficiently that carbon no longer can diffuse through or over the surface of the catalyst particles. It is also possible that the catalyst particles become that much coated with a carbon layer that they cannot absorb more and the nanotube stops growing [21]. The SWCNTs were synthesized in this method shown in Fig. 2.8, in which SWCNTs were catalyzed by a Co–Ni mixture (0.6/0.6 at%), at an oven temperature of 1,200°C. The SWCNTs formed in this case are bundled together by van der Waals forces [36].

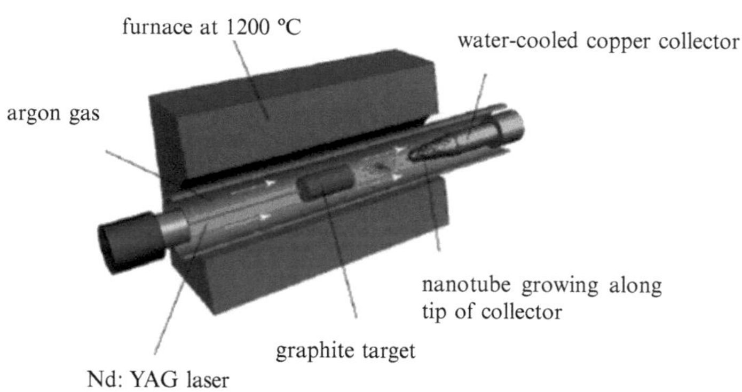

Fig. 2.7 Schematic drawing of a laser ablation apparatus. The pulsed laser evaporates part of the graphite target. The CNT are carried, with the rest of the soot, by an Argon gas flow on to a collector

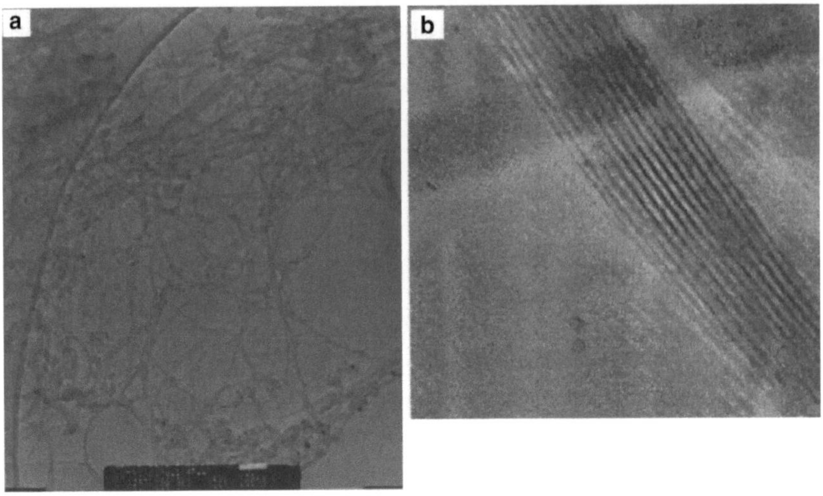

Fig. 2.8 (**a**) Medium-magnification view (scale bar is 100 nm) showing that almost everywhere, tubes are bundled in "multilane highways". Metal particles ranging in diameter from 30 nm down to the resolution limit of the TEM (=0.2 nm). (**b**) High-magnification image of a highway consisting of many adjacent SWCNT. The SWCNT diameters are all about 1 nm, with similar spacing between adjacent nanotubes, which adherer to one another by van der Waals forces. Images are adapted from [16]

2.3 Growth Mechanism of Carbon Nanotubes

So far, the growth mechanism in which nanotubes are formed is still a subject of controversy, and more than one mechanism might be operative during the formation of CNTs. One expected mechanism consists of three steps. Firstly, a precursor to the formation of nanotubes and fullerenes, C_2, is formed on the surface of the metal catalyst particle. Secondly, from this metastable carbide particle, a rod-like carbon is formed rapidly. Finally, there is a slow graphitization of its wall. This mechanism is based on in situ TEM observations [37]. The exact atmospheric conditions will depend on the technique used. The actual growth of the nanotube seems to be similar for all the three synthesis techniques mentioned in Sect. 2.2.

There are several theories on the exact growth mechanism for nanotubes. One theory [14] postulates that metal catalyst particles are floating or are supported on graphite or another substrate. It presumes that the catalyst particles are spherical or pear-shaped, in which case the deposition will take place on only one half of the surface (this is the lower curvature side for the pear-shaped particles). The carbon diffuses along the concentration gradient and precipitates on the opposite half, around and below the bisecting diameter. However, it does not precipitate from the apex of the hemisphere, which accounts for the hollow core that is characteristic of these filaments. Based on supported metals, filaments can form either by "extrusion" (also known as base-growth) (Fig. 2.9a), in which the nanotube grows upwards from the metal particles that remain attached to the substrate, or the particles detach

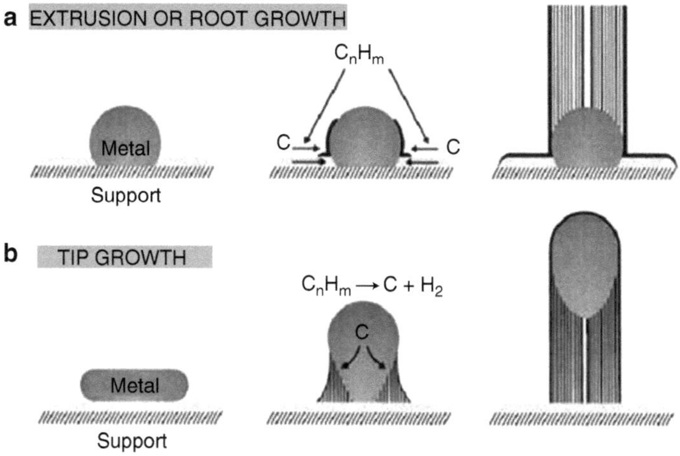

Fig. 2.9 Visualization of a possible carbon nanotube growth mechanism. Images adapted from [26]

and move at the head of the growing nanotube, labeled "tip-growth" (Fig. 2.9b). Depending on the size of the catalyst particles, SWCNTs or MWCNTs are grown. In arc discharge, if no catalyst is present in the graphite, MWCNTs will be grown on the C_2-particles that are formed in the plasma [38].

2.4 Purification of Carbon Nanotubes

In general, the as-prepared materials contain a variety of impurities, such as amorphous carbon, metallic catalyst, and multiwalled graphite, leading to an impediment to more detailed characterization of the properties of SWCNT and their future applications in industry. Therefore, a number of purification procedures have been reported in which the impurities were generally removed by acid oxidation [39–47], hightemperature heat treatment in different atmospheres [42–44], ultrasonic wave [41, 48], or a combination of these methods [49].

In this section, we introduce a simple effective purification method for SWCNTs synthesized by the arcdischarge method. The impurity in the raw material includes amorphous carbon, multiwalled graphite, and metallic catalyst, etc. The method can achieve the controllable removal of these impurities with minimal destruction to SWCNTs.

2.4.1 Purification Method

SWCNTs were synthesized by the arc-discharge method. A graphite rod is used as the discharge anode in the arcdischarge reactor. The graphite rod had a diameter of 6 mm and a hole was drilled along its axis with a diameter of 4 mm and a depth of

60 mm. This hole was filled with a mixture of the graphite and catalysts (Ni and Y powder). The atomic ratio of Ni to Y in the filling was 4.2:1. The argon pressure in the arcdischarge reactor was 200 torr and the gap between the anode and cathode was approximately 5 mm. A discharge voltage of 45–50 V was applied to lead to a direct current of 80–100 A. The cloth-shaped soot was the main product of the arcdischarge process and was used for the further purification investigations.

The as-prepared cloth-shaped soot of 100 mg was firstly dispersed in 300 ml aqueous solution with 1% sodium dodecyl sulfate (SDS) as the surfactant and then ultrasonicated for 2 h. Subsequently, the suspension was filtrated two times with 0.5 μm porous polytetrafluoroethylene filter. Then the solid on the filter was heated in platinum crucibles under the oxygen current with a flow rate of 20 ml min^{-1} at 365°C for 1.5 h. In succession, the sample was washed with concentrated hydrochloric acid to remove metallic catalyst nanoparticles until the solution turned colorless from weak green. The oxidation temperature and the residual metallic catalyst were decided by the TGA and DTA measure, which was done from room temperature to 800°C with a linear heating rate of 10°C min^{-1}. Scanning electron microscopy (SEM) and Raman spectroscopy were used to examine the morphologies and purity of SWCNTs.

2.4.2 Purification Effect

Figure 2.10 shows TGA and DTA curves of the raw cloth-shaped soot and purified SWCNT. A slight weight increase was observed for the raw sample at 180–250°C due to the oxidation of the Ni and Y metal catalysts. There are two stepwise weight loss peaks (352, 428°C), which can be attributed to the oxidation removal of the easy-oxidated carbon component such as amorphous carbon and the SWCNTs successively from the low temperature (Fig. 2.10a).

Only one weight loss peak was observed for the purified SWCNTs (600°C), which corresponds to the SWCNT burning (as Fig. 2.10b). It was observed that the burning temperature of the SWCNTs in the purified sample was higher than that of the SWCNTs in the raw cloth-shaped soot, which could be attributed to the fact that the impurities stuck to SWCNT had been detached in the purified sample. According to the TGA data, the yield of SWCNTs would be very low at temperatures higher than 400°C, whereas the amorphous carbon was not removed effectively at temperatures less than 320°C. Therefore, 365°C is an appropriate temperature to be selected as the oxidation temperature. Moreover, it is well known that DTA is a common method for quantitative analysis, and hence, the residual metal oxide could be estimated to be about 28% and below 1.5%, respectively, for the raw cloth-shaped soot and purified SWCNTs.

Figure 2.11a presents a typical SEM image of the asprepared cloth-shaped soot, which contains a significant amount of amorphous carbon, metallic catalyst and multiwalled graphite nanoparticles. Most of SWCNTs in our sample are several microns to tens of microns long. This provides a possibility for the effective

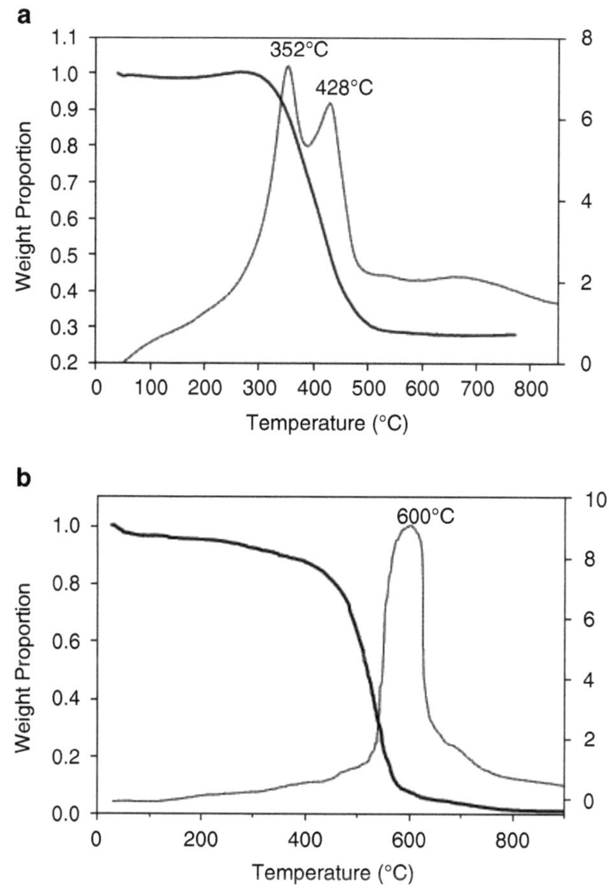

Fig. 2.10 TGA and DTA curves of (**a**) raw cloth-shaped soot and (**b**) purified SWCNTs

elementary separation of the small grainy impurities from SWCNTs by ultrasonication accompanied with filtration before further oxygen oxidation. Figure 2.11b shows the image of purified SWCNTs. It is shown that the impurities can be effectively removed from SWCNTs and a purity of higher than 90% for SWCNTs can be obtained as a result of the purification. The weight of the purified sample is reduced to about 45% of the raw sample.

Raman spectra of cloth-shaped soot and purified SWCNT taken at 1,064 nm are shown in Fig. 2.12. As shown in the figure, the radial-breathing mode (RBM) of the prepared SWCNTs is at 145 and 181 cm^{-1}. The peaks at 1,593 and 1,328 cm^{-1} are known as the G-band and the D-band of SWCNTs respectively. The G-band is the characteristic peak of the crystalline carbon and the D-band represents sp^3 carbon atom, which can originate from the amorphous carbon in the sample or the defects on the SWCNTs. The ratio of the G-band intensity to the D-band intensity (I_G/I_D) can indicate the quantity of the amorphous carbon in the sample or the

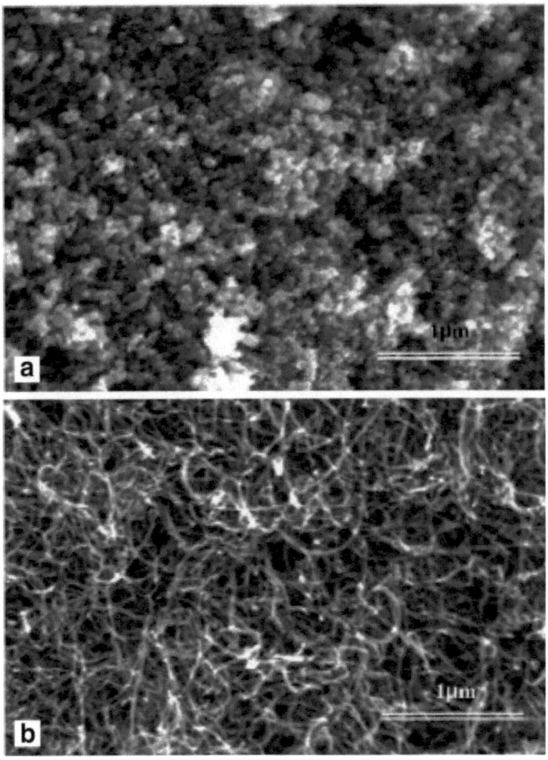

Fig. 2.11 SEM images of (**a**) cloth-shaped soot and (**b**) purified SWCNT

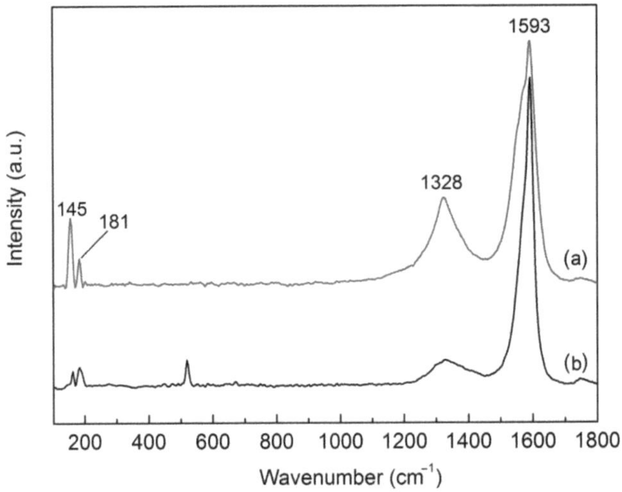

Fig. 2.12 Raman spectra of (**a**) raw cloth-shaped soot and (**b**) purified SWCNTs

defects on the SWCNTs [50]. A significant increase of the ratio of I_G/I_D for the purified SWCNTs in Fig. 2.12 shows a substantial enhancement of the purity and crystallization of SWCNTs after the purification.

The results demonstrate the combination of ultrasonication with the presence of the surfactant and oxidation is an effective method for eliminating the impurities including the amorphous carbon, multiwalled graphite, and metal catalyst etc. The ultrasonication with the presence of the surfactant can make the stuck impurities detached from the SWCNTs, which contributes to their filtering removal and the subsequent oxidation removal of these impurities. The use of the surfactant can cause the SWCNTs to become dispersed homogeneously in the solvent so as to shorten the needed ultrasonication time and can decrease the destruction to SWCNTs caused by the ultrasonication. By exposing the sample to O_2 in the suitable temperature, the impurities including multiwalled graphite and amorphous carbon etc. can be effectively oxidized to be removed while the SWCNTs remain not to be destroyed. Finally, the residual metal catalyst and its oxide can be removed by treating the sample in the concentrated HCl. With these processes, the high-purity minimally destructive SWCNTs can be obtained.

2.5 Conclusion

The past decade has witnessed extensive investigation on the synthesis of SWCNTs, which mainly includes the arc-discharge, CVD, and laser ablation methods. The arc-discharge method can synthesize high-quality nanotubes, but it suffers from the disadvantages such as the plentiful contamination of nanoparticles and other graphitic debris. Luckily, these impurities can be dealt with post-growth chemical purification. The laser ablation synthesis is similar to the arc discharge method in principle, since the optimum background gas and catalyst mix is the same as in the arc discharge process. This might be due to very similar reaction conditions needed, and the reactions probably occur with the same mechanism. The CVD synthesis can produce relatively pure SWCNTs, the diameter and alignment of which can also be more controllable.

The as-prepared CNT raw contains the impurity such as amorphous carbon, metallic catalyst nanoparticles, and large graphitic carbon particles, etc. To remove the various impurities and improve the SWCNT quality, the postsynthesis purification is needed. A simple, efficient, and minimally destructive method for purifying SWCNT has been introduced, in which we combine physical and chemical ways. The results from thermo gravimetric analysis, Raman spectra testing, and SEM observation revealed that a majority of impurities could be effectively removed with the introduced method, leaving SWCNTs with a purity of higher than 90%.

References

1. M.S. Dresselhaus, G. Dresselhaus, P.C. Eklund, *Science of Fullerenes and Carbon Nanotubes* (Academic Press, San Diego, 1996)
2. P.M. Ajayan, T.W. Ebbesen, Rep. Prog. Phys. **60**, 1025 (1997)
3. L.C. Qin, X.L. Zhao, K.R. Hirahara, Y. Miyamoto, Y. Ando, S. Iijima, Nature **408**, 50 (2000)
4. Y. Ando, X. Zhao, K. Hirahara, K. Suenaga, S. Bandow, S. Iijima, Chem. Phys. Lett. **323**, 580 (2000)
5. C. Liu, H.T. Cong, F. Li, P.H. Tan, H.M. Cheng, K. Lu, B.L. Zhou, Carbon **37**, 1865 (1999)
6. S. Iijima, Nature **354**, 56–58 (1991)
7. I.W. Chiang, B.E. Brinson, A.Y. Huang, P.A. Willis, M.J. Bronikowski, J.L. Margrave, R.E. Smalley, R.H. Hauge, J. Phys. Chem. B **105**, 8297 (2001)
8. J.L. Zimmerman, R.K. Bradley, C.B. Huffman, R.H. Hauge, J.L. Margrave, Chem. Mater. **12**, 1361 (2000)
9. M.T. Martinez, M.A. Callejas, A.M. Benito, W.K. Maser, M. Cochet, J.M. Andrés, J. Schreiber, O. Chauvet, J.L.G. Fierro, Chem. Commun. 1000 (2002)
10. H. Hu, B. Zhao, M.E. Itkis, R.C.J. Haddon, J. Phys. Chem. B **107**, 13838 (2003)
11. T.J. Park, S. Banerjee, T. Hemraj-Benny, S.S. Wong, J. Mater. Chem. **16**, 141 (2006)
12. M. Esumi, K. Ishigami, K. Nakajima, A. Sawada, H. Honda, Carbon **34**, 279 (1996)
13. D.S. Bethune, C.H. Kiang, M.S. DeVries, G. Gorman, R. Savoy, J. Vazquez, R. Beyers, Nature. **363**, 605 (1993)
14. Y. Saito, T. Yoshikawa, M. Okuda, N. Fujimoto, K. Sumiyama, K. Suzuki, A. Kasuya, Y. Nishina, J. Phys. Chem. Solids **54**, 1849 (1993)
15. S. Seraphin, D. Zhou, Appl. Phys. Lett **64**, 2087 (1994)
16. T. Guo, P. Nikolaev, A. Thess, D.T. Colbert, R.E. Smalley, Chem. Phys. Lett. **243**, 49 (1995)
17. Y. Saito, K. Nishikubo, K. Kawabata, T. Matsumoto, J. Appl. Phys. **80**, 3062 (1996)
18. C. Journet, W.K. Maser, P. Bernier, A. Loiseau, M. Lamy de la Chapelle, S. Lefrant, P. Deniard, R. Lee, J.E. Fischer, Nature **388**, 756 (1997)
19. Z.J. Shi, Y.F. Lian, X.H. Zhou, Z.N. Gu, Y.G. Zhang, S. Iijima, L.X. Zhou, K.T. Yue, S.L. Zhang, Carbon **37**, 1449 (1999)
20. A. Thess, R. Lee, P. Nikolaev, H.J. Dai, P. Petit, J. Robert, C.H. Xu, Y.H. Lee, S.G. Kim, A.G. Rinzler, D.T. Colbert, G.E. Scuseria, D. Tomanek, J.E. Fischer, R.E. Smalley, Science **273**, 483 (1996)
21. J.V. Holm, M.Sc. thesis (University of Copenhagen, 2007)
22. S.B. Sinnot, R. Andrews, D. Qian, A.M. Rao, Z. Mao, E.C. Dickey, F. Derbyshire, Chem. Phys. Lett. **315**, 25–30 (1999)
23. C. Journet, W.K. Master, P. Bernier, A. Loiseau, M. Lamy, Chapelle, de la. S. Lefrant, P. Deniard, R. Lee, J.E. Fisher, Nature **388**, 756 (1998)
24. Z.F. Ren, Z.P. Huang, J.W. Xu, J.H. Wang, P. Bush, M.P. Siegel, P.N. Provencio, Science **282**, 1105 (1998)
25. Z.F. Ren, Z.P. Huang, D.Z. Wang, J.G. Wen, J.W. Xu, J.H. Wang, L.E. Calvet, J. Chen, J.F. Klemic, M.A. Reed, Appl. Phys. Lett. **75**, 1086 (1999)
26. S.B. Sinnott, R. Andrews, D. Qian, A.M. Rao, Z. Mao, E.C. Dickey, F. Derbyshire, Chem. Phys. Lett. **315**, 25 (1999)
27. A. Fonseca, K. Hernadi, J.B. Nagy, D. Bernaerts, A.A. Lucas, J. Molec. Catal. A: Chem. **107**, 159 (1996)
28. H.M. Cheng, F. Li, G. Su, H.Y. Pan, L.L. He, Appl. Phys. Lett. **72**, 3282 (1998)
29. M. Yudasaka, R. Kikuchi, Y. Ohki, E. Ota, S. Yoshimura, Appl. Phys. Lett. **70**, 1817 (1997)
30. J. Kong, A.M. Cassell, H. Dai, Chem. Phys. Lett. **292**, 567–574 (1998)
31. T. Guo, P. Nikolaev, A. Thess, D.T. Colbert, R.E. Smalley, Chem. Phys. Lett. **243**, 49–54 (1995)
32. B.I. Yakobson, R.E. Smalley, Am. Sci. **84**, 324 (1997)
33. M. Yudasaka, R. Yamada, N. Sensui, T. Wilkins, T. Ichihashi, S. Iijima, J. Phys. Chem. B **103**, 6224 (1999)

34. P.C. Eklund, B.K. Pradhan, U.J. Kim, Q. Xiong, J.E. Fischer, A.D. Friedman, B.C. Holloway, K. Jordan, M.W. Smith, Nano Lett. **2**, 561 (2002)
35. W.K. Maser, E. Munoz, A.M. Benito, M.T. Martinez, G.F. de la Fuente, Y. Maniette, E. Anglaret, J.L. Sauvajol, Chem. Phys. Lett. **292**, 587 (1998)
36. C.D. Scott, S. Arepalli, P. Nikolaev, R.E. Smalley, Appl. Phys. A **72**, 573 (2001)
37. A. Yasuda, N. Kawase, W. Mizutani, J. Phys. Chem. B **106**, 13294 (2002)
38. M. Daenen, R.D. Fouw, B. DeHamers, P.G.A. Janssen, K. Schouteden, M.A.J. Veld, Wondrous world of carbon nanotoube: a review on current carbon nanotube technologies, Project review (Eindhoven University of Technology, 2003)
39. E. Gregan, S.M. Keogh, A. Maguire, T.G. Hedderman, L.O. Neill, G. Chambers, H.J. Byrene, Carbon **42**, 1031 (2004)
40. H.J. Li, L. Feng, L.H. Guan, Z.J. Shi, Z.N. Gu, Solid State Commun. **132**, 219 (2004)
41. K.B. Shelimov, R.O. Esenaliev, A.G. Rinzler, C.B. Huffman R.E. Smalley, Chem. Phys. Lett. **282**, 429 (1998)
42. C.M. Chen, M. Chen, F.C. Leu, S.Y. Hsu, S.C. Wang, S.C. Shi, C.F. Chen, Diam. Relat. Mater. **13**, 1182 (2004)
43. J.G. Wiltshire, A.N. Khlobystov, L.J. Li, S.G. Lyapin, G.A.D. Briggs, R.J. Nicholas, Chem. Phys. Lett. **386**, 239 (2004)
44. K.L. Strong, D.P. Anderson, K. Lafdi, J.N. Kuhn, Carbon **41**, 1477 (2003)
45. H.J. Huang, H. Kajiura, A. Yamada, M. Ata, Chem. Phys. Lett. **356**, 567 (2002)
46. A.C. Dillon, T. Gennet, K.M. Jones, J.L. Alleman, P.A. Parilla, M.J. Heben, Adv. Mater. **11**, 1354 (1999)
47. I.W. Chiang, B.E. Brinson, A.Y. Huang, P.A. Willis, M.J. Bronikowski, J.L. Margrave, R.E. Smalley R.H. Hauge, J. Phys. Chem. B **105**, 8297 (2001)
48. D. Nepal, D.S. Kim, K.E. Geckeler, Carbon **43**, 651 (2005)
49. J.Y. Li, Y.F. Zhang, Physica E **28**, 309 (2005)
50. G.S. Duesberg, I. Loa, M. Burghard, K. Syassen, S. Roth, Phys. Rev. Lett. **85**, 5436 (2000)

Chapter 3
Directed Alignment of Carbon Nanotubes

3.1 Introduction

With extensive long-term researches on carbon nanotubes (CNTs), it is widely recognized that these unique one-dimensional nanostructures possess huge potential as building blocks of next-generation electronic devices and circuits [1, 2]. It is often necessary for CNTs to be assembled in a particular direction when they are used for building various functional electronic and photonic nanodevices. The well-aligned CNT array contributes to achieving high device performance.

Several techniques have been adopted to align the CNTs Zhang et al. showed that applying an electric field during the chemical vapor deposition (CVD) growth of single-walled carbon nanotubes (SWCNTs) could make them bridge two electrodes orientedly [3]. Lewenstein et al. reported a chemical patterning method for selective deposition of CNTs in which the alignment and assembly were achieved by a charge interaction between CNTs and patterned charge layer [4]. Filter deposition from a suspension in strong magnetic fields was used to prepare a thick film of oriented SWCNTs [5]. A well-designed, steady laminar flow on a sheet could also be utilized to produce the shear forces for aligning disordered and entangled SWCNTs into bulk ordered material [6]. The electrical-field-assisted alignment had also been demonstrated as an effective method for the aligned deposition of CNTs, in which CNTs dispersed in a suspension are moved and aligned between electrodes under a DC or AC electric field [7–13].

In the studies, the dispersive, directed CNT arrays or films are highly desired for the high-performance or special functional device application due to the fact that one-dimensional characteristics of CNTs will be remained in the configuration. Although the CNTs could be aligned in specific direction approximately with the aforementioned many methods, the CNTs in prepared directed arrays or films often had an entanglement and lacked a good dispersivity, which might be attributed to the confinement of the used alignment method or the lack of the pretreatment of CNTs. The entanglement between deposited CNTs will degrade the alignment quality and be adverse to the achievement of high-performance devices. In addition, the density and thickness control of aligned CNTs is also important for meeting the need of different device constructions, especially in the case of the large-area alignment. In this chapter, an electric field-assisted alignment method and a Langmuir–Blodgett (LB) alignment technique will be introduced respectively, for the alignment of the SWCNTs prefunctionalized with octadecylamine (ODA). For the electric filed assisted alignment method, the dispersive, directed SWCNT arrays can be obtained, and with the LB alignment method, large-area aligned SWCNTs with controllable density and thickness can be prepared.

3.2 AC Electric Field-Assisted Alignment of Carbon Nanotubes

3.2.1 AC Dielectrophoresis Method of SWCNTs

SWCNTs with an average diameter of 1.4 nm synthesized by arc discharge method were used. To improve the dispersivity and solubility of SWCNTs in organic solvents, the following surface modification was conducted to SWCNTs. Firstly, purified SWCNTs were dispersed in H_2SO_4–HNO_3(3:1) and kept at 80°C for 30 min in the reflux system for introducing the carboxyl at their opening ends and the defect locations on their walls. Then the SWCNTs were mixed with ODA and N, N'-dicyclohexylcarbodiimide(DCC) and reacted at 60 or 90°C for 24 h, to graft the ODA onto SWCNTs. Afterward, the excess ODA and DCC were removed by rinsing with ethanol for five times (10 min sonication at 40 kHz). After filtration, the product was dried in vacuum at room temperature. The obtained SWCNTs were ultrasonically dissolved in the chloroform or isopropanol to form a dispersive, stable SWCNT solution [14].

The parallel Au electrode patterns were made using standard UV lithography and lift-off technique on n-type silicon wafers with a 500-nm thermally oxidized layer. The electrodes were 100 nm in thickness, 10 μm in length, and had a gap of 1 μm. Both AC and DC electric fields were utilized to align the SWCNTs. The applied AC bias held a voltage of 10 V peak-to-peak and a frequency of 5 MHz. The applied DC voltage was 10 V. The alignment was achieved by dripping a droplet of SWCNT solution (typical SWCNTs concentration = $0.5\,\mu g\,ml^{-1}$) on the electric field biased electrodes. The electric field was not switched off until the solution was evaporated up so as to left aligned SWCNTs between the electrodes on the wafer. Finally, the ODA grafted on the surface of SWCNTs was removed by exposing SWCNTs in UV light with 194 nm wavelength for 60 min, which was produced from a high-pressure Hg lamp. The UV irradiation caused the alkyl chains to be decomposed and evaporated up while SWCNTs remained on the substrate [15, 16].

3.2.2 Solubility and Dispersivity of Surface Modified SWCNTs

Figure 3.1 shows the scanning electron microscope (SEM) image and infrared absorption spectrogram for the modified SWCNTs. The obtained SWCNTs are highly purified as observed in Fig. 3.1a. The stretching vibration of C = O bond in

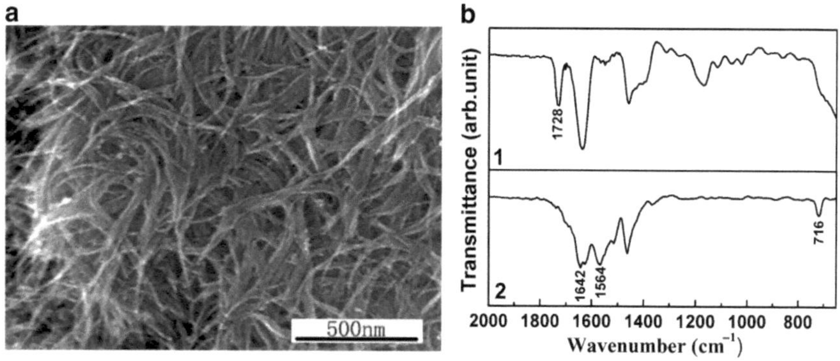

Fig. 3.1 SEM image and infrared spectrograms of the treated SWCNTs. (**a**) SEM photograph of SWCNTs after the surface modification. (**b**) infrared spectrograms of the SWCNTs in different stages 1: after reacting with the mixed acid; 2: after the surface modification with ODA

carboxyl (1728 cm^{-1}) is observed after the mixed acid treatment (see Fig. 3.1b), which shows that the carboxyl is introduced onto SWCNTs. After reacted with ODA and DCC, the stretching vibration mode of C = O bond in carboxyl disappears, while the peak at 1642 cm^{-1} assigned to the stretching vibration of C = O bond in amide appears. Besides, the peaks at 1,564 and 716 cm^{-1} respectively corresponding to the in-plane and out-plane bending vibration of N–H bond are also observed in the spectrum. These results indicate that the SWCNTs are grafted by ODA. The reaction can be expressed as

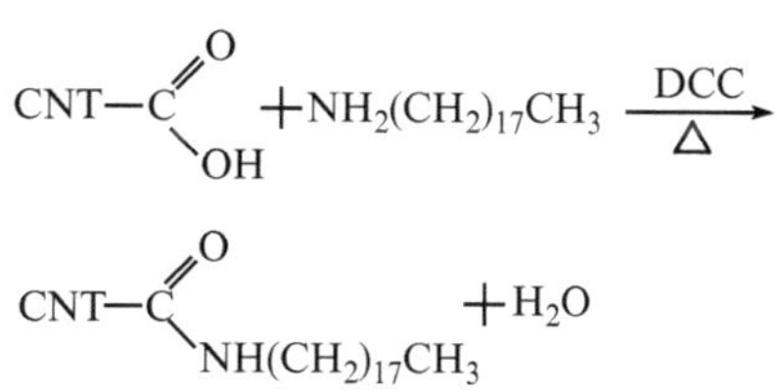

It is noticed that the reaction temperature has a significant influence to the extent of reaction. The mole percent of grafted ODA to C in SWCNTs can be expressed as $\delta = \frac{(m_2 - m_1)/M_{\mathrm{ODA}}}{m_1/M_{\mathrm{C}}} \times 100\%$, where m_1 and m_2 are respectively the mass of SWCNTs before and after the surface modification and M_{ODA} and M_{C} are respectively the mole mass of ODA and C. According to this equation, the mole percent, δ, of the grafted ODA is calculated as about 1.01% and 2.13% respectively at 60 and 90°C, which is slightly lower than the value of 3–4% [17] or 3.6–8.0% [18] in the previous studies. Thus, with the elevation of reaction temperature, the reaction extent is enhanced and the quantity of grafted ODA is increased. This is also confirmed by the infrared spectrogram analysis. As shown in the Fig. 3.2, the peaks pertaining to ODA, e.g. the peak at 716 cm^{-1}, for the sample prepared at 90°C are stronger than those for the sample with the same concentration prepared at 60°C.

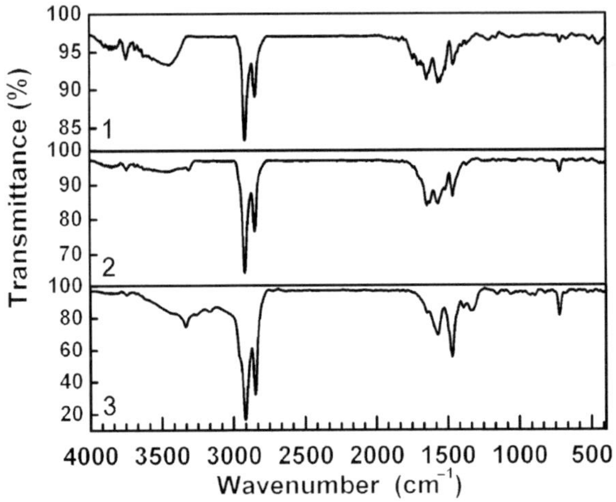

Fig. 3.2 Infrared spectrograms of the samples with the same concentration modified in different temperatures. 1: modified at 60°C; 2: modified at 90°C; 3: pure octadecylamine

The solubility and dispersivity of SWCNTs is greatly enhanced by the surface modification treatment. In contrast to the as-prepared SWCNTs, which are insoluble in organic solvents, the modified SWCNTs manifest a substantial solubility and good dispersivity in many kinds of organic solvents such as the chloroform, dichloromethane, CS_2 etc. No precipitation was observed for the SWCNTs in these solutions with the concentration up to 300 μg ml^{-1} upon prolonged standing. In addition, the surface modification may contribute to the weakening of the mutual attractive force between SWCNTs, causing exfoliation of the SWCNTs bundles to give individual nanotubes [10].

3.2.3 Directed Alignment Effect of SWCNTs

The SWCNTs are highly aligned along the direction of the electric flux after the alignment. Figure 3.3a, b show the scanning electron microscope (SEM) images of aligned SWCNTs obtained by applying an AC voltage of 10 V at 5 MHz with chloroform and isopropanol as the solvents respectively. It indicates that the periodical and dispersive SWCNT array is formed. The SWCNTs are approximately vertical to metal electrodes. The average spaces between neighboring SWCNTs are about 200 nm and 150 nm respectively. It is found that the volatility of the solvents has a significant influence on the alignment effect of the SWCNTs. The shorter the evaporation time, the fewer the contaminants around the electrodes. As shown in Fig. 3.3a, b, much more contaminants, presumably amorphous carbon particles, are found to distribute around the electrodes when the premodified SWCNTs are aligned in the isopropanol under AC electric field, compared with the alignment

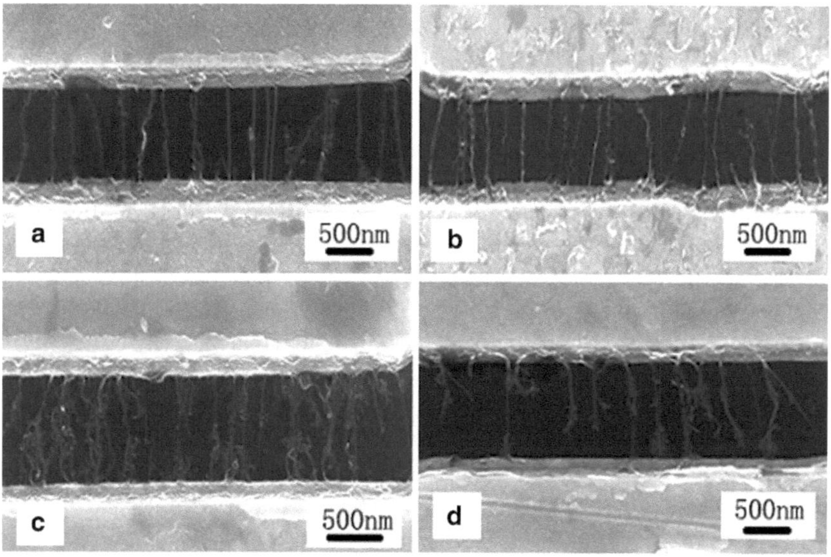

Fig. 3.3 SEM images of the samples aligned in different conditions with the $0.5\,\mu g\,ml^{-1}$ concentration of SWCNT solution. (**a**) Aligned in the chloroform with an AC bias after the surface modification. (**b**) Aligned in the isopropanol with an AC bias after the surface modification. (**c**) Aligned in the chloroform with an AC bias without the surface modification. (**d**) Aligned in the chloroform with a DC bias after the surface modification

result in the chloroform with the higher volatility. Making use of this purification effect, the SWCNTs can be extracted from background contaminants by aligning in the high-volatility solvents.

To account for the formation of aligned array and the influence of the solution volatility, one needs to learn about the mechanism for electric field-assisted alignment. In the alignment processes, two forces can affect the movement of SWCNTs toward the electrodes. One is the electrophoretic force, which comes from the interaction of the electric field and non-zero net charges [19]. The other is the dielectrophoretic force, which is produced by the interaction of a nonuniform electric field with the electric field-induced dipole. In an AC electric field, the movement due to the electrophoresis force is oscillatory with zero time-average and only dielectrophoretic force causes the movement of the SWCNTs. The dielectrophoretic force drives SWCNTs to move toward regions of high electric field along the electric field gradient, thereby to be deposited around the electrodes where the electric field strength is highest. Besides the translational movement, the SWCNTs also experience a rotary movement during the alignment process, because the induced dipoles also bear a torque force at the same time. Thus, the SWCNTs are ultimately aligned in the direction of applied electric field by adjusting themselves along tube axis parallel to the electric field direction, as shown in the SEM images. Similarly, the contaminants in the AC electric field also endure the effect of the dielectrophoretic force. Due to the structural anisotropy of SWCNTs, the induced dipole moment in the direction parallel to the length of the tube axis is much stronger than that in the perpendicular direction, while the nanoparticles have uniform dipole moment irrespective of the axes. The strong polarization along their length direction makes SWCNTs move fast toward metal electrodes, while the contaminants respond with slow movement to the applied electric field and take longer time to reach the electrodes. Due to that the high-volatility solvent is evaporated up rapidly before the contaminants reach the electrodes, the deposition of contaminants is effectively prevented.

As a comparison, the same solution concentration of SWCNTs without surface modification ultrasonicated in the chloroform were also aligned with an AC electric field. Different from the alignment results with the surface modification, for this time the neighboring SWCNTs are entangled together, although visible orientation is also observed (Fig. 3.3c). The difference shows that the surface modification contributes to obviating the entanglement of SWCNTs and to forming the dispersively aligned SWCNT array. It is believed that the improvement of the solubility and dispersivity of SWCNTs after the surface modification is responsible for this. Besides, the repulsion between net charges on SWCNTs may also contribute to align SWCNTs dispersively and periodically, since some net charges are demonstrated to be held on the modified SWCNTs by aligning SWCNTs under the DC electric field. It is found that when a DC bias is used to align SWCNTs, although directed alignment is also obtained, some of them are only attached to the positively biased electrode (the upper electrode in Fig. 3.3d) and cannot make the reliable electrical connection with both electrodes. This is not observed for the surface-unmodified samples. Therefore, it can be deduced that some negative charges should be held on SWCNTs after surface modification, which interact with the DC electric field and cause SWCNTs to be attracted toward the positively biased electrode.

It is also found that the density of the SWCNTs aligned under the DC electric field is much lower than that of the SWCNTs aligned under AC electric field of high frequency. A few nanoparticles are observed to be deposited around the aligned SWCNTs (see Fig. 3.3d). One of possible illustrations for this is that the modified SWCNTs have a larger dielectric constant under the AC electric field of high frequency than under the DC electric field. Hence, the SWCNTs are deposited across the electrodes more rapidly under the AC electric field of high frequency, which makes two electrodes conduct and slows down the deposition of the contaminants. This is in good agreement with earlier reports that aligning SWCNTs in high-frequency AC electric field has a function of purifying the raw SWCNTs [20, 21].

3.2.4 Electrical Characteristics of Aligned SWCNTs

The electrical transport characteristics of aligned SWCNT array are also studied in the experiment. It is noticed that modified SWCNTs cannot form the good electrical contact with metal electrodes. As shown in Fig. 3.4a, the zero-bias resistance of the as-prepared SWCNT array with 11 SWCNTs aligned between electrodes is up to 1.03 MΩ, because the ODA grafted on the surface of SWCNTs cumbers the direct contact of SWCNTs and electrodes and causes large contact resistance. After

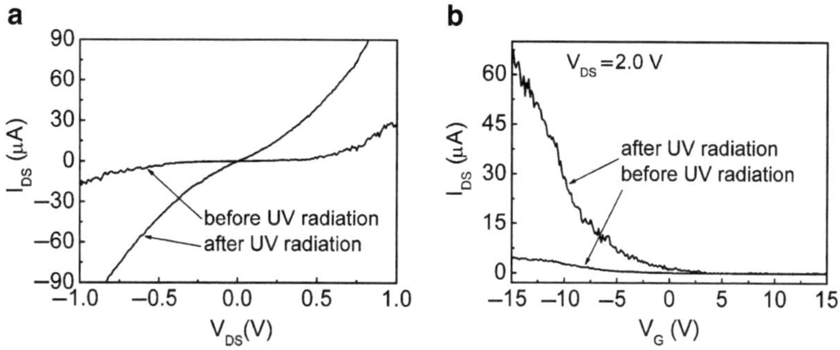

Fig. 3.4 Electrical transport characteristics of aligned SWCNT array before and after UV radiation. (**a**) Current–voltage characteristics of the SWCNT array with 11 SWCNTs aligned between electrodes before and after UV radiation. (**b**) Transfer characteristics of the CNTFET with 14 SWCNTs as the conduction channels before and after UV radiation

the devices are exposed to UV irradiation, the zero-bias resistance is decreased to $0.019\,\mathrm{M\Omega}$. This can be attributed to that the alkyl chains are decomposed and evaporated up under UV radiation [22, 23]. From the results, we found that although the UV radiation decreased the contact resistance largely, the resistance of average single SWCNT was still large relative to the resistance of ideal ballistic SWCNT ($6.5\,\mathrm{k\Omega}$). Due to the fact that the Au electrode is difficult to chemically react with SWCNTs under the relatively weak UV energy, it is considered that the SWCNTs were coupled to the electrode by the weak van der Waals force after the UV irradiation. This caused a large contact resistance to remain after irradiation. UV radiation also enhances the performance of fabricated CNTFET.

Figure 3.4b shows the transfer characteristics of a CNTFET before and after UV radiation, which was prepared by burning off the metallic SWNTs in an array using an electrical breakdown method [24, 25]. SEM observation shows that 14 SWCNTs remain to act as the conduction channels of this CNTFET after the burning-off process. As shown in Fig. 3.4b, after the UV radiation, the on current of CNTFET at $V_G = -15\,\mathrm{V}$ and $V_{DS} = 2.0\,\mathrm{V}$ varies from 4.6 to $66.8\,\mu\mathrm{A}$, which is increased by more than one order of magnitude, and the transconductance in the linear region rises from 0.52 to $7.52\,\mu\mathrm{S}$. The on–off current ratio at $V_{DS} = 2.0\,\mathrm{V}$ amounts to $10^4 - 10^5$, while before removing the ODA it is only about 2.0×10^3. The substantial enhancement of the electrical characteristics clearly indicates that a much improved interface between SWNTs and electrodes is achieved by using the UV radiation. Therefore, the UV radiation has a function of reversing the surface modification.

3.3 Langmuir–Blodgett Technique for the Alignment of Carbon Nanotubes

In previous studies, the LB technique had been used to incorporate SWCNTs in LB organic multilayers [26]. The multilayers consisting of cadmium behenate (CdBe) layers and SWCNT layers could be fabricated by this technique. The dependency of the packed density, coverage status and layering properties of SWCNTs and the compression pressure in the LB technique had been studied. Recently, the excellent electrical, thermal and mechanical properties and promising applications for the directed film consisting of SWCNT itself have induced great research interest on it. In this section, we mainly introduce the application of the LB technique in preparing controllable-density, highlyordered SWCNTs on the substrates of glass, silicon, quartz, etc. [27].

3.3.1 Langmuir–Blodgett Method

The purified SWCNTs were treated in mixed acid (H_2SO_4–HNO_3 in volume 3:1) at 50°C for 30 min to introduce carboxylic groups onto the SWCNTs. The obtained products were reacted with thionyl chloride to introduce acyl chloride groups onto the SWCNTs, and then reacted with octadecylamine (ODA) to cause them to become grafted onto SWCNTs [28]. The functionalized SWCNTs have substantial solubility in chloroform, CS_2 and various aromatic solvents. No precipitation was observed for the SWCNTs in these solutions with the concentration up to 300 $\mu g\, ml^{-1}$ upon prolonged standing. The results of optical absorption characterization for the treated SWCNTs were almost the same as that of those reported for nonfunctionalized SWCNTs [29], indicating that the electronic structure of SWCNTs is not subject to significant change after the pretreatments and chemical functionalization.

The LB instrument used in the work was a KSV 5,000. A chloroform solution of SWCNTs (1.7 $mg\, ml^{-1}$) was spread at the air–water interface in the Langmuir trough at an interval of 20 s. After the solvent was evaporated up, the barrier began to move to compress the SWCNT film at a constant speed. The compression pressures (surface pressures) were monitored continuously in this process. Two deposition methods of the film including horizontal dipping and vertical dipping were used in the studies, the schematical diagrams of which are given in Fig. 3.5b, c. For horizontal dipping, when the surface pressure was 30 $mN\, m^{-1}$ the substrates were slowly brought down until they were in contact with the floating monolayer. After the film deposition, the film-coated substrates were inclined with a gradient of approximately 5° in order to ensure that the subphase water was removed without disturbing the transferred film and then were lifted. For vertical dipping, the end of the hydrophilic substrates was immersed vertically in the pure water in the trough in advance. After the SWCNT solution was distributed on the water surface, the barrier began to work. When the surface pressure reached 30 $mN\, m^{-1}$ the substrates were dipped down at 15, 10 and 2 mm min^{-1} speeds to transfer SWCNTs onto them. All the films were prepared at 20°C.

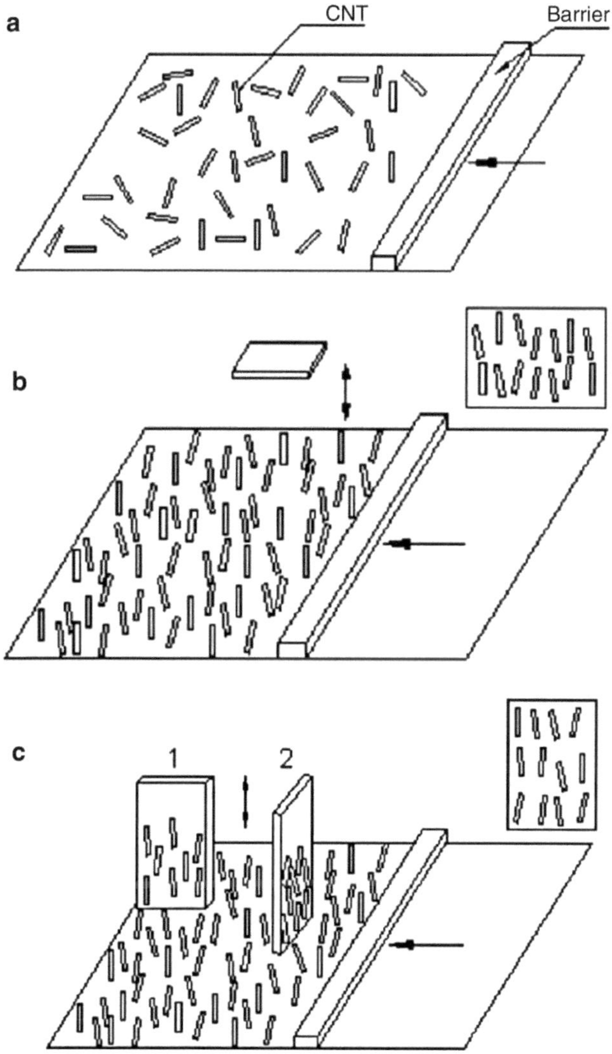

Fig. 3.5 Scheme of compression-induced orientation in the Langmuir trough by barrier compression (**a**), the preparation of films by the horizontal dipping (**b**), and vertical dipping methods (**c**). Case 1 and case 2 in (**c**) represent the different placement ways perpendicular and parallel to the barrier for the substrate, respectively

3.3.2 Alignment Effect of Langmuir–Blodgett Method

Figure 3.6 shows smooth surface pressure–area (p–A) isotherms for 1 ml SWCNTs chloroform solution ($1.7\,\mathrm{mg\,ml^{-1}}$) at different compression speeds. It should be noted here that the actual trough area was used as the abscissa instead of area

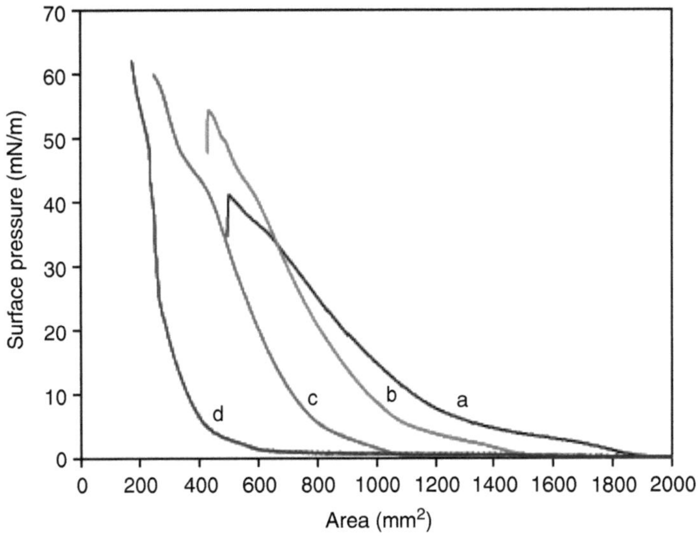

Fig. 3.6 Surface pressure–area (p–A) isotherms of SWCNTs at different compression speeds (**a**) 30, (**b**) 20, (**c**) 10, and (**d**) 4 mm min^{-1}

per molecule because these SWCNTs are a mixture of tubes with different diameters and lengths. It is observed that when the barrier speed reduces from 30 to 4 mm min^{-1} the collapse surface pressure increases from about 40 to 60 mN m^{-1}. As shown in Fig. 3.6 (d), a steep rise for the surface pressure in the range of 10–45 mN m^{-1} and a high collapse surface pressure is observed when the barrier speed is 4 mm min^{-1}. Therefore, a barrier speed of 4 mm min^{-1} and a surface pressure of 30 mN m^{-1} were adopted in the vertical dipping deposition.

Figure 3.7 shows SEM images of the deposited films on the silicon by horizontal dipping method with different compression pressures. It is observed that at low surface pressure the SWCNTs are sparsely packed (Fig. 3.7a, b). In contrast, if the surface pressure is too high, the SWCNT film on the subphase will be packed too tightly and will be easy to collapse, resulting in agglomerated regions, which contain several-layer SWCNTs (Fig. 3.7d). When the surface pressure is about 30 mN m^{-1}, it is observed that SWCNTs are directedly aligned well (Fig. 3.7c). For the horizontal-dipping deposition method, the compressioninduced orientation is considered to be the only orientation mechanism of the SWCNT film.

In the fabrication of the SWCNT LB film, the chemical modification is necessary since SWCNTs are different from traditional amphiphilic molecules like fatty acids. The carboxylic and amide groups introduced onto SWCNTs by the chemical treatments will play an important role in the formation of the SWCNT film on the water surface. These groups are located at the nanotube ends and the defect spots of nanotube sidewalls [30]. These introduced groups will help to introduce hydrophilicity into the nanotubes, which is originally very hydrophobic before the chemical treatments, thus making it possible for SWCNTs to spread uniformly on

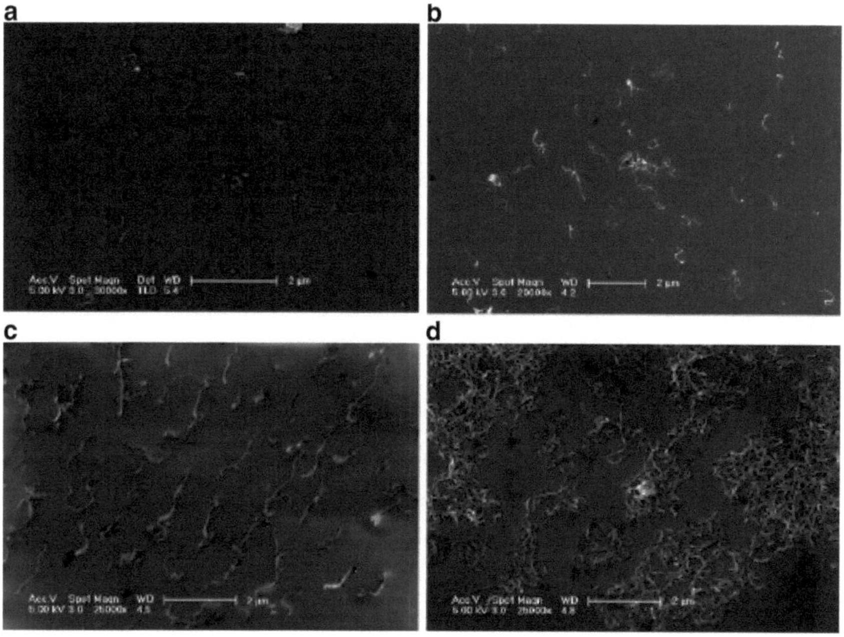

Fig. 3.7 SEM images of SWCNTs films prepared by horizontally dipping at different surface pressures (**a**) 0 mN m^{-1}, (**b**) 10 mN m^{-1}, (**c**) 30 mN m^{-1}, and (**d**) 50 mN m^{-1}

the water surface. This is similar to the acquirement of the monolayer of α-helical copolyglutamates with long side chains [31], in which rigid rod-shape molecules can spread on the water surface with an appropriate chemical modification.

Figure 3.8 shows AFM images of SWCNTs films on silicon by vertical dipping, the arrow corresponding to the dipping direction. It is clear that nanotubes are preferentially oriented in the dipping direction. The measure of the surface profiles of AFM images (Fig. 3.8c, d) shows that the tube height distribution ranges mostly from 5 to 9 nm, indicating that most tubes are in small bundles the length of which ranges from a few hundred nanometers to several micrometers. It is observed that the quantity of SWCNTs deposited onto the silicon would increase and the orientation effect would also be better as the dipping speed decreases. During the transferring process, it is most important to keep a constant surface pressure. At a too high dipping speed, the film does not have enough time to flow and be deposited well (Fig. 3.8a), and many defects are formed as the increased pressure caused by the high dipping speed makes the film fracture. The actual deposition rate will be limited by the response time of the compression system. Therefore, a low deposition rate was chosen in the study.

Different from the horizontal-dipping deposition method, both the flow-induced orientation and compressioninduced orientation will be expected to work in the vertical-dipping deposition method. The flow-induced orientation is due to the fact that the flow of the water along the dipping direction on the substrate when the

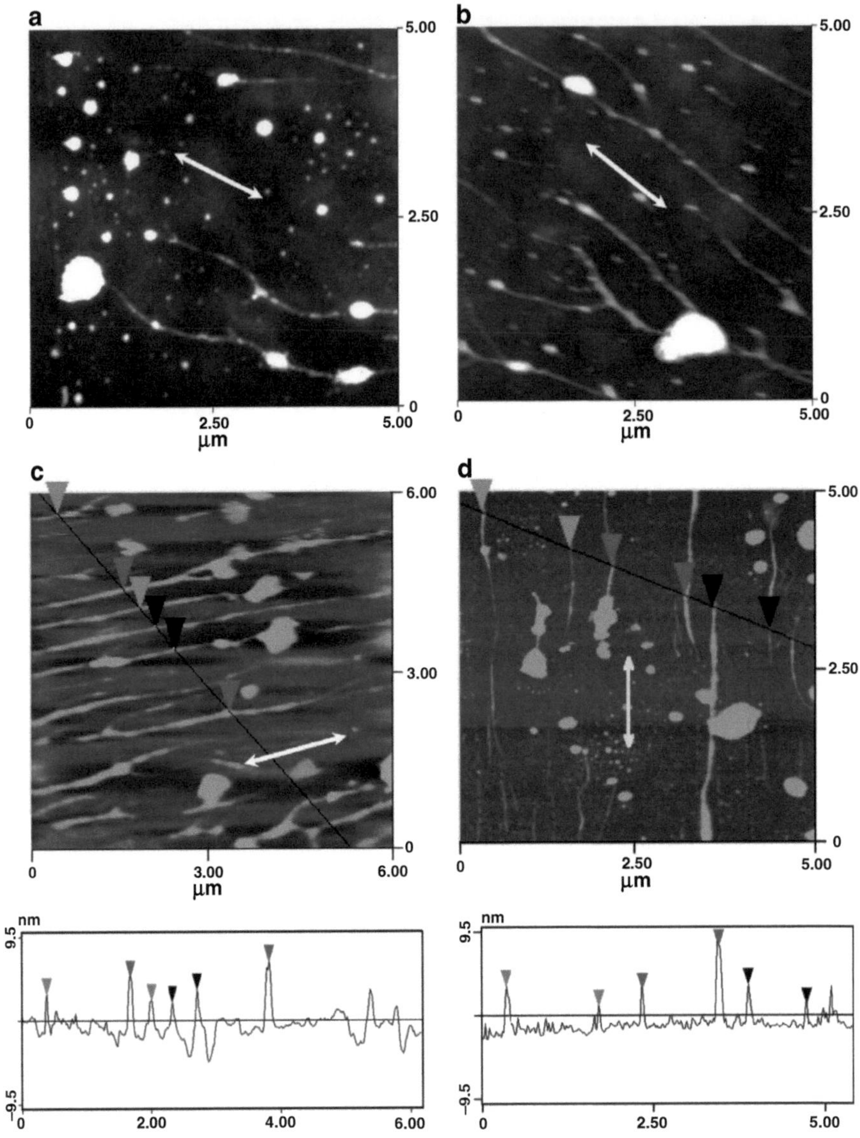

Fig. 3.8 AFM images of SWCNTs films transferred by the vertical dipping at (**a**) 15 mm min^{-1}, (**b**) 5 mm min^{-1}, and (**c, d**) 2 mm min^{-1} dipping speeds; A section analysis along the black line is also given in (**c**) and (**d**). For (**a–c**), the substrate is placed to be perpendicular to the barrier; for (**d**) the substrate is placed to be parallel to the barrier. The arrows in the figures show the dipping direction

substrate is dipped enables SWCNTs aligned along the dipping direction. To illuminate the interplay between compressioninduced and flow-induced orientation, the effect of the different placement ways of the substrate on the SWCNT orientation

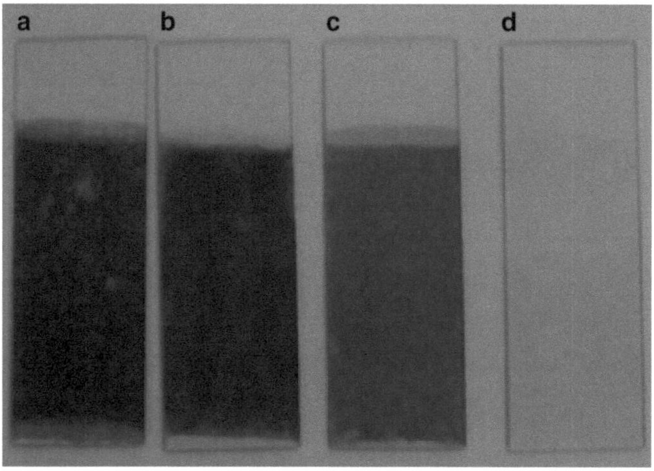

Fig. 3.9 Photographs of SWCNT films prepared by the vertical dipping at 30 mN min^{-1} dipping speed. The layer number of SWCNT films is (**a**) 132, (**b**) 82, (**c**) 26, and (**d**) 4, respectively

is studied. For case 1 (Fig. 3.8a–c), in which the substrate was placed to be perpendicular to the barrier, the flow-induced orientation direction is identical to the compression-induced orientation direction, which would be advantageous to the orientation effect. For case 2 (Fig. 3.8d), in which the substrate was parallel to the barrier, the flow-induced orientation would be expected to work against the compression-induced orientation. However, the actual experiments had shown little difference of the orientation results between case 1 and case 2. More detailed studies are needed to understand this.

The large area SWCNTs films on the glass substrate (size: 20 mm × 60 mm × 2 mm) could be obtained (see Fig. 3.9) by a layer-by-layer fashion (132, 82, 26 and 4 layers respectively), which allows one to readily control the film thickness. In order to obtain high-quality multilayer films, it is important that the initial layers are well formed since the faults in these layers may be replicated in the subsequently deposited layers. To ensure a good adhesion, the initial layers were deposited on the substrate at low immersion and withdrawal rate (2 mm min^{-1}). When a few layers had been formed (e.g. eight layers) the rate was increased to a higher value. To improve the film deposition, it is often appropriate to temporarily halt the dipping process after an upstroke and wait until the transferred monolayer is completely dry before continuing the deposition cycle.

3.3.3 Polarized Raman Spectroscopy of Aligned SWCNTs

The orientation of SWCNTs in LB film was further investigated by means of polarized Raman spectroscopy. Figure 3.10 shows the polarized Raman spectra of SWCNTs LB film with 17 layers deposited at 30 mN m^{-1} on hydrophilic quartz.

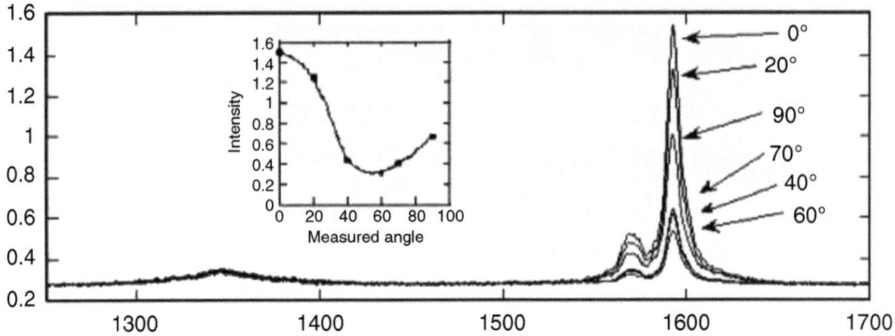

Fig. 3.10 Polarized Raman Spectra of the SWCNTs LB film with 17 layers on hydrophilic quartz taken at 514.5 nm (2.1 eV). *Inset*: Intensity changes of the G-band at 1,590 cm^{-1} with the measured angle from $\theta = 0°$ to $\theta = 90°$. Images are adapted from the [23]

Raman peaks at 1,590 cm^{-1} are attributed to the tangential stretching G-band modes. The peak intensities varied with the measured angle (h) between the polarization direction and the nanotubes axis. Particularly, the experimental intensities of G-band modes exhibit a characteristic minimum near $\theta = 60°$ in excellent agreement with the theory value of SWCNTs [24]. The inset in Fig. 3.10 shows the intensity change of the G-band at 1,590 cm^{-1} as a function of the measured angle from $\theta = 0°$ to $\theta = 90°$. Since the laser beam was polarized parallel to the dipping direction of the substrate ($\theta = 0°$), the aforementioned results indicated that the orientation of SWCNTs tended in the dipping direction. This is consistent with the result of AFM observations.

3.3.4 Alignment Mechanism of Langmuir–Blodgett Method

For the LB films of SWCNTs, two mechanisms can influence its orientation: compression-induced and flow-induced. When compressed, SWCNTs with large aspect ratios floating on water surface are expected to be orientated with their axis parallel to the barrier of Langmuir trough. This is the orientation mechanism of the horizontal dipping deposition. For the vertically dipping deposition, an additional mechanism, flow-induced orientation, comes into effect due to the relative flow of the water on the substrate. In the flow process, SWCNTs will be drained from the water surface. This drainage is a result of the adhesion between the transferred monolayer and the substrate, which makes the transferred material disengage the water film and aligned along the flow direction. The actual alignment process may be complicated by other factors such as the length, diameter, and entanglement of SWCNTs and the further study on the mechanism is still ongoing.

3.4 Conclusion

Dispersively directed carbon nanotube array can be obtained between metal electrodes by AC dielectrophoresis of surface modified SWCNTs. The graft of ODA onto SWCNTs achieves the well-dispersed SWCNT solution and effectively obviates the entanglement of deposited SWCNTs. It is indicated that the quality of directed placement can be enhanced by using a high-volatility solvent and an AC electric field of high frequency. The good electrical contact between the aligned SWCNTs and the metal electrodes is obtained by adopting an UV-light radiation method for removing the grafted ODA on the aligned SWCNTs.

With the LB technique, large-area aligned SWCNTs with controllable density and thickness can be prepared. The alignment can be achieved on the substrates including glass, silicon, quartz, etc. Chemically functionalized SWCNTs are found to posses good surface spreading properties that contributed to the formation of stable Langmuir films on water surface. The overall direction of the carbon nanotubes alignment can be controlled both by compression and dipping process. The LB deposition of SWCNTs can be performed in a layer-by-layer fashion up to several hundred layers, allowing to readily control the film thickness.

The prepared dispersed, directed SWNCT array is expected to have an application in the fabrication of high-performance functional devices. The solution-based 'bottom-up' assembly approach would enable facile integration of diverse nanowires building blocks with more complex structure and would provide a pathway for fabricate nanodevices.

References

1. X. Sun, R. Li, B. Stansfield, J.P. Dodelet, S. De'silets, Chem. Phys. Lett. **394**, 266 (2004)
2. K.L. Strong, D.P. Anderson, K. Lafdi, J.N. Kuhn, Carbon **41**, 1477 (2003)
3. Y. Zhang, A. Chang, J. Cao, Q. Wang, W. Kim, Y. Li, N. Morris, E. Yenlimez, J. Kong, H. Dai, Appl. Phys. Lett. **79**, 3155 (2001)
4. H. Dai, J.H. Hafner, A.G. Rinzler, D.T. Colbert, Nature **384**, 147 (1996)
5. J.E. Fischer, W. Zhou, J. Vavro, M.C. Llaguno, C. Guthy, R. Haggenmueller, J. Appl. Phys. **93**, 2157 (2003)
6. H. Yu, W. Zhou, G. Ning, Q. Zhang, G. Luo, F. Wei, Carbon **43**, 2215 (2005)
7. K. Philip, C.M. Lieber, Science **286**, 2148 (1999)
8. J. Park, A.N. Pasupathy, J.I. Goldsmith, C. Chang, Y. Yaish, J.R. Petta, Nature **417**, 722 (2002)
9. J. Kong, N.R. Franklin, C.W. Zhou, M.G. Chapline, S. Peng, K. Cho, Science **287**, 622 (2000)
10. R.H. Baughman, C.X. Cui, A.A. Zakhidov, Z. Iqbal, J.N. Barisci, G.M. Spinks Science **284**, 1340 (1999)
11. L.M. Dai, P.G. He, S.N. Li, Nanotechnology **14**, 1081 (2003)
12. J.Q. Li, Q. Zhang, D.J. Yang, J.Z. Tian, Carbon **42**, 2263 (2004)
13. N. Peng, Q. Zhang, J.Q. Li, N.Y. Liu, J. Appl. Phys. **100**, 024309 (2006)
14. C.X. Chen, Y.F. Zhang, J. Phys. D: Appl. Phys. **39**, 172 (2006)
15. R.H. Baughman, A.A. Zakhidov, W.A. de Heer, Science **297**, 787 (2002)
16. Y. Nakayama, S. Akita, Synth. Met. **117**, 207 (2001)
17. S.M. Huang, M. Woodson, R. Smalley, J. Liu, Nano Lett. **4**, 1025 (2004)
18. Y.L. Li, I.A. Kinloch, A.H. Windle, Science **304**, 276 (2004)

19. H.H. Gommans, J.W. Alldredge, H. Tashiro, J. Park, J. Magnuson, A.G. Rinzler, J. Appl. Phys. **88**, 2508 (2000)
20. X.Q. Chen, T. Saito, H. Yamada, K. Matsushige, Appl. Phys. Lett. **78**, 3714 (2001)
21. M.S. Kumar, T.H. Kim, S.H. Lee, S.M. Song, J.W. Yang, K.S. Nahm, E.K. Suh, Chem. Phys. Lett. **383**, 235 (2004)
22. Y.Z. Guo, M. Mitsuishi, T. Miyashita, Macromolecules **34**, 3548 (2001)
23. Y.Z. Guo, J. Wu, Y.F. Zhang, Chem. Phys. Lett. **362**, 314 (2002)
24. P.G. Collins, M.S. Arnold Ph. Avouris, Science **292**, 706 (2001)
25. P.G. Collions, M. Hersam, M. Arnold, R. Martel Ph. Avouris, Phys. Rev. Lett. **86**, 3128 (2001)
26. T.D. Luccio, F. Antolini, P. Aversa, G. Scalia, Tapfer, L. Carbon **42**, 1119 (2004)
27. J.Y. Li, Y.F. Zhang, Carbon **45**, 493 (2007)
28. J. Chen, M.A. Hamon, H. Hu, Y. Chen, A.M. Rao, P.C. Eklund, Science **282**, 95 (1998)
29. I.W. Chiang, B.E. Brinson, A.Y. Huang, P.A. Willis, M.J. Bronikowshi, J.L. Margrave, J. Phys. Chem. B **105**, 8297 (2001)
30. M.A. Hamon, H. Hu, P. Bhowmik, S. Niyogi, B. Zhao, M.E. Itkis, Chem. Phys. Lett. **347**, 8 (2001)
31. G. Duda, A.J. Schouten, T. Arndt, G. Lieser, G.F. Schmidt, C. Bubeck, Thin Solid Films **159**, 221 (1988)

Chapter 4
Ultrasonic Nanowelding Technology Between Carbon Nanotubes and Metal Electrodes

4.1 Introduction

Carbon nanotubes (CNTs) exhibit a great prospect as a nanoscale building block for future nanoelectronics due to their unique one-dimensional nanostructure and properties [1]. To explore their potential in various domains, an essential prerequisite is to build reliable interconnections between the CNTs and the external electrical circuits and mechanical systems. To address this need, various chemical and physical processes have been explored to build such interconnections. For example, Burghard et al. reported "controlled adsorption of CNTs on chemically modified electrodes" for interconnection of CNTs [2,3]. However, a stronger bonding instead of a weak chemical adsorption is mandatory for constructing reliable nanodevices. Ruoff et al. showed that focused electron beam in a scanning electron microscope (SEM) can be used to deposit a small amount of hydrocarbon contamination so as to attach nanotubes on an AFM tip [4,5]. Such a "spot welding" technique has also been used for connecting CNTs and polysilicon surface electrically and mechanically [6]. Madsen et al. presented an in situ method for highly conductive attachment of multiwalled carbon nanotubes (MWCNTs) onto microelectrodes by depositing a gold–carbon composite using a focused electron beam system [7]. Although robust contacts can be obtained by the these methods, limited access to a focused electron beam system and the small-scale spot-treatment nature prevent their large-scale industrial applications. To meet the needs of future large-scale applications, simpler, less capital intensive and more scalable processes are highly desirable.

In this chapter, we introduce a novel ultrasonic nanowelding process, with which one can fabricate reliable bonding between single-wall carbon nanotubes (SWCNTs) and metal electrodes. Contacts formed by the present process are found to have low contact resistance and good long-term stability and mechanical strength. A low resistance of around $15\,k\Omega$ for a $1-\mu$m-long metallic SWCNT at room temperature can be achieved [8]. After nanowelding, the effective Schottky barrier height between semiconducting SWCNT and Ti electrode is as low as $\sim$6.6 meV in the ON-state and the barrier width is $\sim$0.9 nm at $V_g = 0$. The performance of the CNT field-effect transistors (FETs) fabricated by this method has also been demonstrated to have greatly improved.

4.2 Conventional Fabrication Method of Carbon Nanotube/Metal Contacts

To study their electrical properties, SWCNTs are required to connect to external electrodes or systems [9–13]. This section describes how the conventional method attaches leads to individual nanotubes or nanotube bundles to conduct the electrical measure.

To attach leads to the nanotubes, we start with a degenerately doped Si wafer that has a thermal oxide grown on the surface. Optical lithography is used to etch holes in the oxide and define Al contacts to the degenerately doped Si, which is used as a gate. In this procedure, the wafer is first covered with photoresist. Exposure to UV light through a mask chemically changes the resist. The pattern defined by the mask is then developed to produce holes in the resist layer. The wafer is then immersed in HF, which removes the exposed oxide. Finally, Al is evaporated onto the wafer, and then the unwanted metal is removed by lift-off process. These steps are shown in Fig. 4.1a–c [12].

Starting with this wafer with back gate contacts, making electrical contact to the nanotubes is then done in either of two ways. In the first method, leads are first defined using the optical lithography or electron-beam lithography (EBL) to make an

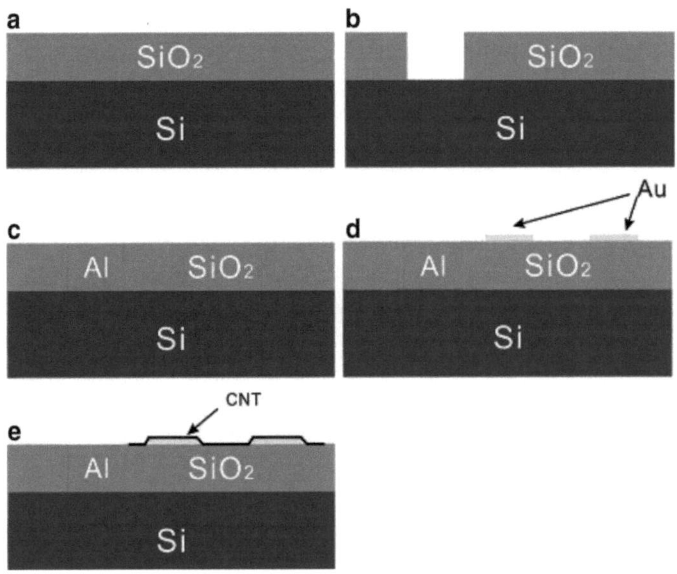

Fig. 4.1 Fabrication steps for a nanotube device. (**a**) Starting substrate consisting of a degenerately doped Si wafer with 1 μm of thermally grown oxide. The degenerately doped Si is used as a gate electrode. (**b**) Optical lithography and an HF etch create windows in the oxide layer. (**c**) Contact to the gate is made by evaporating Al through the windows. (**d**) Leads are defined using a combination of optical and ebeam lithography. (**e**) Nanotubes are deposited on the leads to form the completed device. Images adapted from [12]

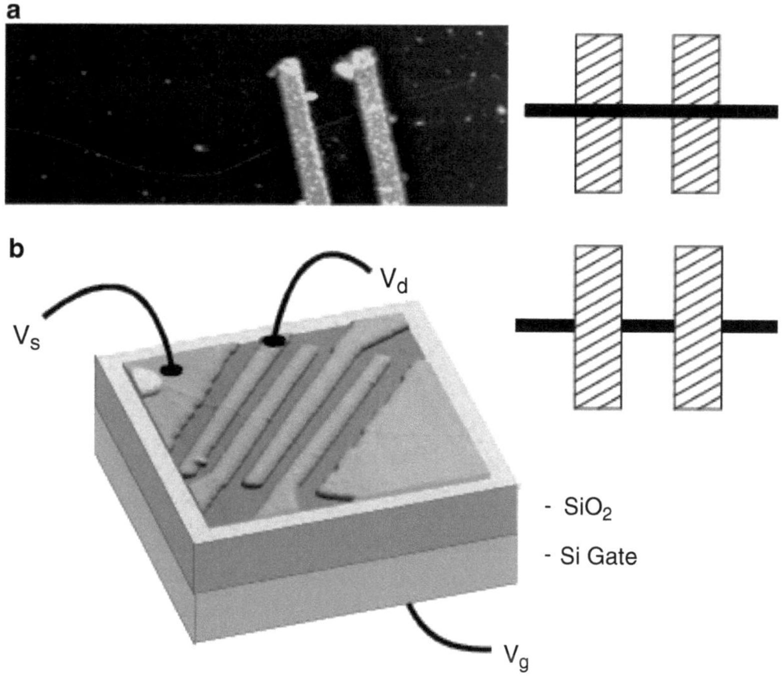

Fig. 4.2 (**a**) A device made by depositing the tubes on top of the leads. (**b**) A device made by evaporating the leads on top of the nanotubes. The source and drain leads are shown as well as the lead to the back gate. Images adapted from [12]

array of electrodes. Tubes are deposited on top of the leads, as is shown schematically in Fig. 4.1d, e. Promising candidate devices are then found by using a probe station to test the electrical properties. These devices are then examined with an AFM to determine the number and size of the nanotubes/bundles bridging the contacts. Figure 4.2a shows a device made using this technique. In the second method, nanotubes are deposited or grown on an oxidized Si wafer that has predefined alignment marks in addition to the back gate contacts. These alignment marks consist of $\sim$1 μm Au squares that were formed in a prior lithography step. A suitable tube or bundle is located with an AFM, and its position is noted relative to the alignment marks. A resist polymer (either PMMA or PMMA/MAA) is then spun on the sample and leads to the nanotubes are defined with EBL. A completed device is shown in Fig. 4.2b.

In either method, once the leads have been attached, the device is mounted in a chip package using colloidal silver paint. Electrical contact between the bonding pads on the chip and the package is made by 25-μm-diameter Al bonding wire. Finally, the package can then be inserted into standard 16-pin sockets and is ready for electrical measurements.

4.3 Ultrasonic Nanowelding Technology of Carbon Nanotubes to Metal Electrodes

Although the contact between CNTs and metal electrodes can be formed by the aforementioned processes, the contacts obtained by these methods suffered from the problems of the large contact resistance, poor process reliability, or the complex process procedure etc. When barely depositing CNTs on the top of high-work function ($\Phi > 5\,\text{eV}$) metal electrodes, a high contact resistances ($>1\,\text{M}\Omega$) between CNTs and electrodes could be caused. Tersoff [14] explained that this high contact resistance might come from the inhibition to the conduction between the metal and CNT by the Bloch symmetry of the electrons or the unique electronic structure of the contact system, resulting in a weak electronic coupling at the Fermi surface. Besides, the high contact resistance can be attributed to the weak van der Waals bonds between metal and CNTs when CNTs are simply laid on the surface of metal electrodes, which will cause some separation between metal and CNT so as to produce a serious barrier for the conduction of electrons [14, 15]. Moreover, the simple placement of CNTs on the electrodes will also result in the lack of reliability in the devices. For the mentioned second case placing the metal electrodes on the CNTs, the complex process procedure is needed, which is disadvantageous to the large-scale device fabrication. The contact resistance for the devices without optimizing the contact is still high. These disadvantage factors prevent further practical applications of CNT-based devices. To achieve CNT-based devices with the small size, high speed, and low-power dissipation, the sturdy and low-resistance contact is mandatory [16].

4.3.1 Process of Ultrasonic Nanowelding

SWCNTs with an average diameter of 1.6 nm synthesized by an arc discharge process were used in this work. SWCNTs were purified and subjected to chain-scission in a 3:1 mixture of concentrated sulfuric acid and nitric acid at 80°C for 30 min in a reflux system. The undried SWCNTs were filtrated with isopropanol to remove the water in the sample. Then the SWCNTs were ultrasonicated in isopropanol and centrifuged at 15,000 rpm for 2.5 h to further separate large SWCNT bundles from single SWCNTs. The obtained supernatants at concentrations of several micrograms per milliliter were ultrasonically treated for about 20 h to sufficiently disperse SWCNTs. Parallel titanium electrode pairs were patterned by optical lithography on n-type silicon wafer, which was thermally coated with a 150-nm thick silicon dioxide layer. The electrode pairs were 300 nm thick, 40 μm wide, and 40 μm long, and were separated by a gap of 1 μm. SWCNTs were deposited onto the electrodes by simply performing an unassisted deposition of SWCNTs in suspension or by an AC electric-field assisted alignment process [17, 18]. A droplet of the SWCNT suspension was introduced onto the patterned wafers. After the deposition or electric field alignment process, SWCNTs were laid over the electrode pair. The number of

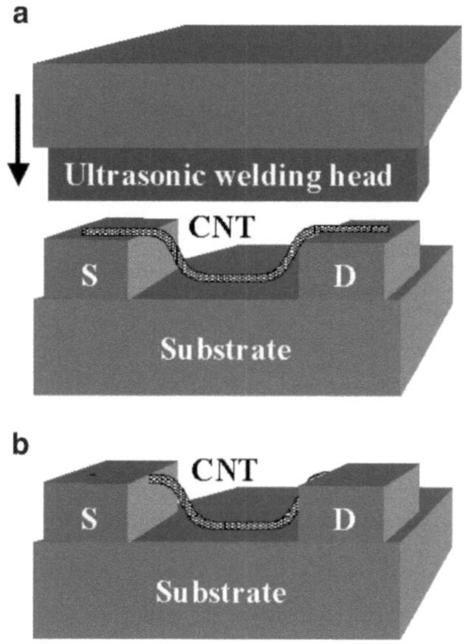

Fig. 4.3 Schematics of the ultrasonic nanowelding process. (**a**) Before welding and (**b**) after welding

the SWCNTs bridging the electrode pair can be readily controlled by varying the concentration of the suspension or the parameters of applied AC electric-field such as the bias magnitude and frequency. The as-prepared samples were then ready for ultrasonic bonding [8].

Ultrasonic nanowelding was carried out in a FB-128 ultrasonic wire bonder. The structure and the details on how this wire bonder works can be found in [19]. An Al_2O_3 single crystal with a 2,500 μm^2 circular pressing surface and a RMS roughness of 0.2 nm was mounted onto the bonder to act as the welding head. Figure 4.3 shows the schematic of the ultrasonic nanowelding process. A clamping force of 78.4 mN was applied to press the welding head against the nanotube and electrodes. At the same time an ultrasonic vibration with a frequency of 60 kHz was applied to the welding head through an ultrasonic transducer (Fig. 4.3a). The ultrasonic energy was transferred to the bonding interface through the ultrasonic welding head. Thus the ends of SWCNTs and electrodes were welded together under the combined action of the ultrasonic energy and a clamping force (Fig. 4.3b). To investigate the effects of ultrasonic energy on the bonding process, a serial of different ultrasonic powers was applied. The welding process was carried out at room temperature for a duration of 0.2 s.

4.3.2 *Effect of Ultrasonic Nanowelding*

Figure 4.4a, b show typical scanning electron microscopy (SEM) images of a SWCNT bridging two electrodes before and after welding with an ultrasonic power of 0.07 W. It can be observed that a SWCNT was hanging over the two electrodes

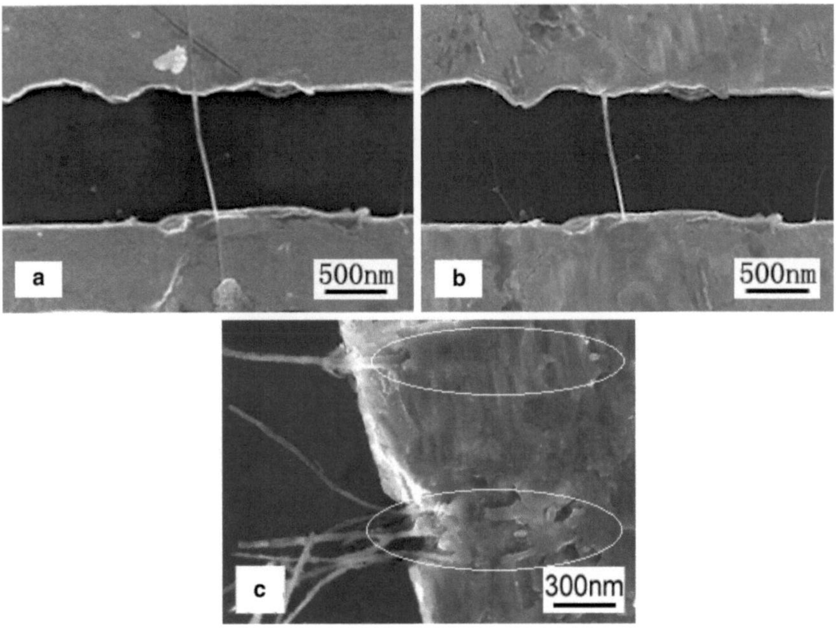

Fig. 4.4 (**a, b**) SEM images of an individual SWCNT bridging the Ti electrodes before and after the ultrasonic nanowelding with an ultrasonic power of 0.07 W respectively. (**c**) SEM image of multiple SWCNTs welded onto the electrode with an ultrasonic power of 0.16 W. The marked regions show the weld junctions

before welding. The morphology of the nanotube on the two electrodes can be clearly observed. After welding, the ends of SWCNT was embedded into the electrodes and the nanotube morphology was now almost invisible on the electrodes. Scratches caused by the friction between the ultrasonic welding head and the electrodes can also be observed on the electrode surfaces (Fig. 4.4b). Figure 4.4c shows that multiple nanotubes were welded onto the titanium electrodes under pressing as well as ultrasonic treatment.

In addition, the ultrasonic nanowelding can also be used for welding bulk quantities of SWCNTs onto metal electrodes. Figure 4.5 shows SEM images of a 2,500 μm^2 electrode surface with nanotubes welded onto it. As shown in Fig. 4.5a, the ultrasonic nanowelding produces an apparent directional welding zone due to the directional vibration of the welding head. In the nonwelded zone, the SWCNTs loosely stay on the surface of the metal electrode (Fig. 4.5b). Inside the welding zone, the SWCNTs and metal substrate are bonded together to form a new weld surface. Almost all SWCNTs are embedded and welded into the Ti electrode and only the ends of some short SWCNT protruded from the welded substrate can be observed (see Fig. 4.5c).

The mechanism behind the ultrasonic nanowelding is proposed as follows. During the ultrasonic nanowelding, the high-frequency ultrasonic energy softens the metal and causes plastic deformation of the metal under the clamping stress

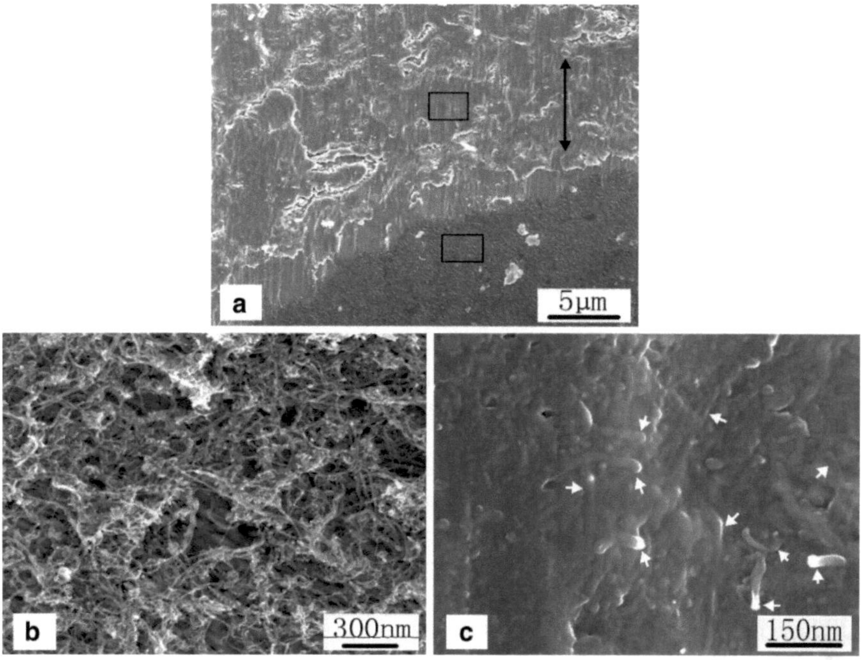

Fig. 4.5 SEM images of the electrode surface after welding of bulk quantities of SWCNTs onto Ti electrode at an ultrasonic power of 0.21 W. (**a**) Surface topography of the welded (upside) and nonwelded (downside) zone. The arrow represents the direction of the ultrasonic vibration motion. (**b, c**) Zoom-in views of a local region (indicated by rectangles) in the unwelded and welded zone respectively. The *arrows* in (**c**) show that some short ends of SWCNTs protrude from the welded substrate

because of the "acoustic softening effect" [19, 20]. Thus, the nanosized SWCNTs with a one-dimensional structure have shown to be "embeddable" and weldable into the metal electrodes.

4.3.3 Nature of Nanowelded Carbon Nanotube/Metal Contacts

To examine the nature of the welded bonds, microarea X-ray photoelectron spectroscopy (XPS) and X-ray diffraction (XRD) analysis were used to study the welded zone. The data are shown in Fig. 4.5a. The XPS analysis reveals that the C 1s corc level spectrum (Fig. 4.6a, curve 1) of the region shows a single peak at 284.5 eV when the ultrasonic power is lower than 0.14 W. This peak corresponds to the C–C bond of the original SWCNTs. This result indicates that no detectable chemical reaction happens between the nanotube and the metal electrode at low ultrasonic power. When a higher power (>0.14 W) was used, a new peak appeared at around 280.8 eV (Fig. 4.6a, curve 2). This peak can be attributed to C–Ti bond in titanium carbide (Ti_xC), which indicates that Ti_xC compounds could have been formed in

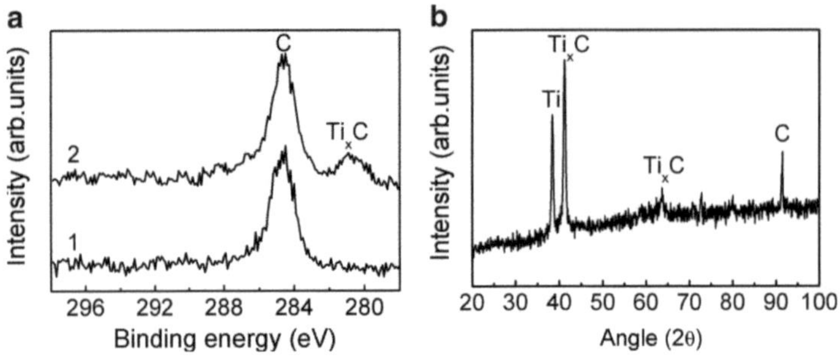

Fig. 4.6 (**a**) C 1s core level XPS spectra of the weld zone with SWCNTs welded onto a Ti electrode at the ultrasonic power of 0.07 W (1) and 0.21 W (2) respectively. The spot diameter of X-ray beam is 20 μm. (**b**) Microarea XRD diffractogram of the weld zone after bonding at the power of 0.19 W. The spot diameter of X-ray beam is 10 μm

this process. The microarea XRD analysis result is also in good agreement with this interpretation. An XRD peak at $2\theta \approx 65.2°$, which represents a titanium carbide (Ti_xC) phase, was observed when the ultrasonic power was increased to above 0.14 W. When ultrasonic power was increased to 0.19 W, two peaks of Ti_xC phase are observed in the XRD pattern at $2\theta \approx 41.5°$ and $\approx 63.7°$, as shown in Fig. 4.6b. All these results indicate that when the ultrasonic power is higher than a threshold value, SWCNTs do chemically react with the titanium electrode to form titanium carbide at the welding junction. More comprehensive and in-depth study is needed to substantiate this observation.

4.4 Characteristics of Nanowelded Carbon Nanotube/Metal Contacts

4.4.1 Electrical Characteristics of Nanowelded Metallic SWCNTs

To evaluate the electrical performance of the contact, two-terminal (2t-) resistance tests have been carried out on the nanotube welded across the electrode pairs. To simplify the interpretation of the results, "metallic nanotubes" rather than "semiconducting nanotubes" were used in the two-terminal resistance tests. The metallic and semiconducting nanotubes were identified and differentiated by their different dependences on gate voltages applied to the silicon substrate. In the experiment, it was found that both individual SWCNTs and SWCNT bundles comprised of a small number of SWCNTs were present in the isopropanol suspension, although many SWCNT bundles can be dispersed into individual SWCNTs as shown in Fig. 4.7. As a result, both individual SWCNTs and SWCNT bundles might bridge the electrodes after the deposition of SWCNTs. In our experiment, the individual SWCNTs bridging the electrodes were selected for testing by measuring their diameters using atomic force microscopy (AFM), hence excluding the SWCNT bundles. The structure was considered to be an individual SWCNT when the measured AFM diameter was smaller than 2.2 nm. Figure 4.8 shows two examples of

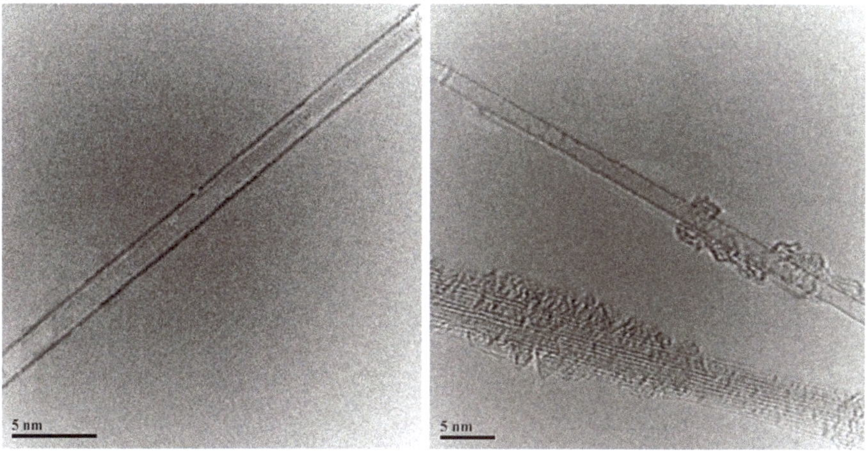

Fig. 4.7 HRTEM images of individual SWCNTs in isopropanol after the dispersion

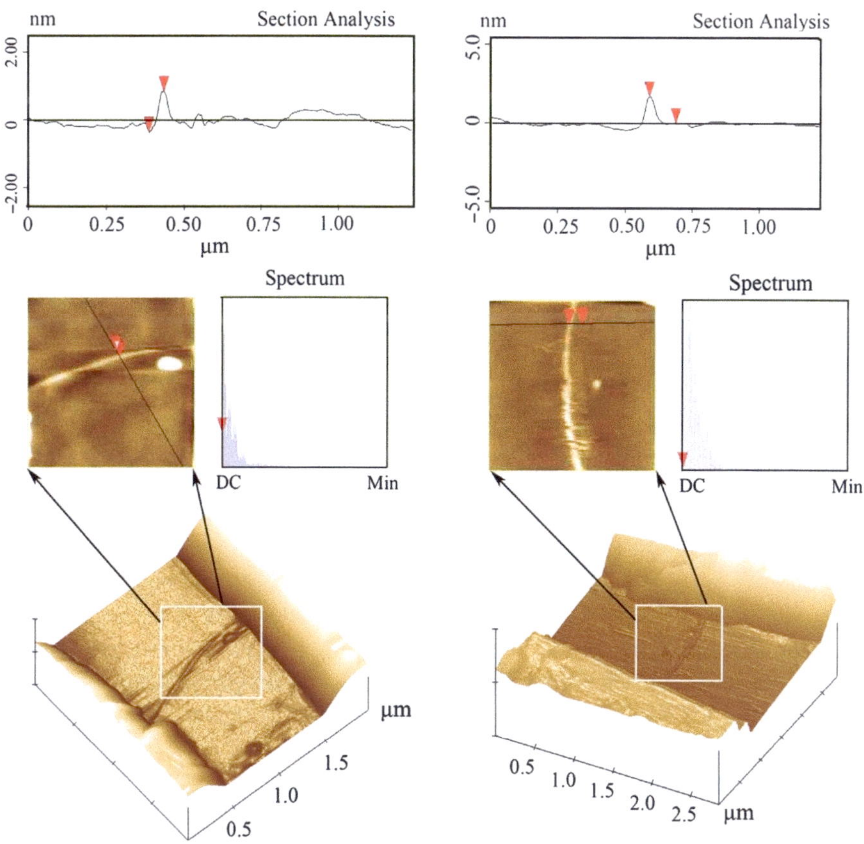

Fig. 4.8 Two typical samples for measuring the diameters of SWCNTs by AFM

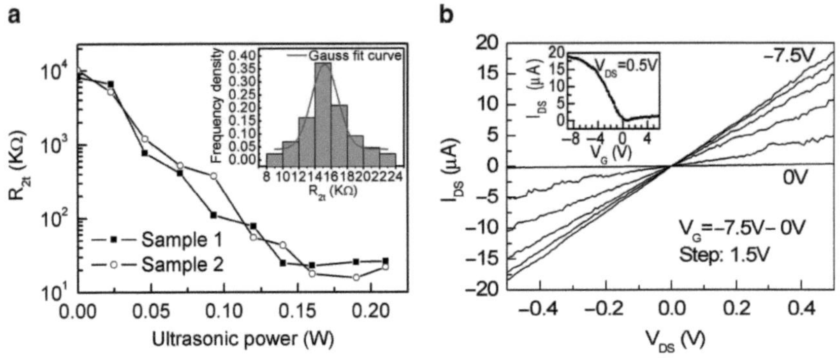

Fig. 4.9 Electrical performances of individual metallic and semiconducting SWCNTs after the ultrasonic nanowelding. (**a**) Two terminal (2t-) resistances as a function of the ultrasonic power for two metallic nanotube–Ti contacts. Before the ultrasonic nanowelding, the 2t-resistances of sample 1 and 2 were 49.4 and 55.8 MΩ respectively; The ultrasonic power of 0 W in the figure represents the ends of the SWCNT pressed against the electrodes by pure clamping force without ultrasonic power applied. *Inset*: the statistical analysis and distribution fit of the obtained lowest 2t-resistances for 43 samples after nanowelding. (**b**) Output characteristic of the CNTFET with individual semiconducting SWCNT as conduction channel. *Inset*: transfer characteristic curve

the measure to an individual SWCNTs. Before the ultrasonic nanowelding, the 2t-resistances of individual metallic SWCNTs were in the order of tens of megaohms at room temperature. After ultrasonic nanowelding with a power of 0.16–0.19 W, the 2t-resistances decreased to a narrow range of 8–24 kΩ. To further evaluate the effects of the ultrasonic power on the nanowelding fabrication, samples were successively welded with a serial of increasing ultrasonic powers ranging from 0 to 0.21 W, at intervals of about 0.023 W. Figure 4.9a shows the dependence of the 2t-resistances as a function of the ultrasonic power for two representative devices. Before welding, the 2t-resistances of these two samples were 49.4 and 55.8 MΩ respectively. When the ends of the nanotubes lying on the metal electrodes were pressed against the electrodes by a pure clamping force with no ultrasonic power applied, the 2t-resistances of two devices dropped to 8.1 and 10.2 MΩ respectively. It was demonstrated that the 2t-resistances decreased with the increase of the applied ultrasonic powers. As the ultrasonic power increased to 0.15 W, the 2t-resistances decreased by more than three orders of magnitude, and then saturated at the low values of about 22 and 15 KΩ respectively. In the experiments, the 2t-resistances of 43 samples were measured and statistically analyzed. It was found that the obtained lowest resistances for these samples mainly concentrated around 15 KΩ and approximately conformed to the Gauss distribution (inset in Fig. 4.9a), indicating that the formed electrical contacts were reliable. Due to the fact that the ultrasonic bonding will not affect the intrinsic resistance of the SWCNTs between the source and drain electrodes, all decrease of the 2t-resistance can be attributed to the decrease of the contact resistance. For the decrease ratio of contact resistance, it can be expressed as

$$k = \frac{R_1 - R_{\mathrm{CNT}}}{R_2 - R_{\mathrm{CNT}}}, \tag{4.1}$$

where R_1 and R_2 are the 2t-resistance before and after the ultrasonic bonding respectively, and R_{CNT} is the intrinsic resistance of the SWCNTs between the source and drain electrodes. From the expression (4.1), the decrease ratio of contact resistance will be expected to be higher than the decrease ratio of 2t-resistance. All of these results indicate that the contact resistance can be greatly reduced by the ultrasonic bonding [15]. The contacts were also found to be stable in air in a long-term basis and showed little change after storage for three months at room temperature in laboratory atmosphere.

The remarkable decrease of 2t-resistance could be attributed to the following reasons. One, SWCNTs are embedded into the metal electrodes after the ultrasonic bonding, since the "acoustic softening effect" of the high-frequency ultrasonic energy can soften the metal and allow nanoscale SWCNT to embed into the metal more easily under the clamping press [19, 20]. Thus, the effective contact area between SWCNTs and electrodes is increased, resulting in a better electrical contact [21]. Another possible reason is that some structural impairment or lattice defects at the ends of the SWCNTs would be produced during the ultrasonic bonding. These impairments or defects would increase the electron scatter at the junction and contribute to the breaking of the Bloch symmetry of the electrons (i.e. wave vector conservation) between SWCNTs and the metal electrodes, thereby allowing an easier electrical conduction between them [14]. Besides, since the high-level ultrasonic treatment may provide the activation energy for interdiffusion [19,22,23], the interface reaction between SWCNTs and the metal electrodes such as the formation of titanium carbides could occur, which will decrease their contact resistance. Further study is required in order to validate these proposed mechanisms.

4.4.2 Electrical Characteristics of Nanowelded Semiconducting SWCNTs

Characteristics of the bonds between semiconducting SWCNTs and metal electrodes, created using this welding technique have also been studied. Several back-gate CNTFETs were prepared by welding individual semiconducting SWCNTs onto electrodes with ultrasonic power in the range of 0.16–0.19 W. The output and transfer characteristics of a typical CNTFET are shown in Fig. 4.9b. The transistor shows p-type characteristics, with a low threshold voltage of –0.3 V. As shown in Fig. 4.9b, I_{DS}–V_{DS} curves are highly linear and symmetrical for $-0.5\,V < V_{DS} < 0.5\,V$. The bonded semiconducting SWCNT exhibits a large ON current $I_{ON\_sat} = 18.9\,\mu A$ and a high linear ON conductance of $G_{ON} = 0.25 \times 4e^2/h$ ($R_{ON} = 26.4\,k\Omega$) at $V_G = -9.0\,V$, indicating the low-resistance electrical contact in the ON state. The obtained CNTFET shows a high transconductance of $g_{m|V\,ds=0.5\,V} = 3.6\,\mu S$ in the linear region of $-4.5\,V < V_G < -1.0\,V$ (inset in Fig. 4.9 b), while before the ultrasonic nanowelding it is only in an order of 10^{-9} S. This transconductance is the highest among present solid-state back-gate individual nanotube FETs [18, 21, 24–27], which is a little higher than that of the Ti-contacted top-gated

SiO_2/SWCNT-FETs ($g_{m|V_{ds}=-1.0\,V} = 3.25\,\mu S$) [28] and is higher by one order of magnitude than that of the Ti-contacted back-gated SiO_2/SWCNT-FETs using annealing method to form low-resistance contacts ($g_{m|V_{ds}=0.8\,V} = 0.2\,\mu S$) [21, 24]. The inverse subthreshold slope (S) of our devices is around 180 mVdec^{-1}, which is much lower than that of those devices in the same manner without ultrasonic nanowelding ($S \sim 400$ mVdec^{-1}) [29] and is identical to that of those devices with bottom gates and Al_2O_3 dielectrics ($S \sim 180$ mVdec^{-1}) [25]. Therefore, the on–off performance of the devices was well improved after nanowelding. A large on/off current ratio of $10^6 - 10^7$ has also been achieved. For the obvious performance improvements in the CNTFET, the significant decrease in the contact resistance is clearly an important cause. Another possible explanation can be attributed to the change in the transistor configuration from the "side-bonded" nanotube-metal contact to the "end-bonded" nanotube-carbide contact due to the formation of titanium carbides, which enhanced the modulation effect of the gate [24].

4.4.3 Barrier of Nanowelded Contacts

Temperature dependence of 2t-transport experiments was conducted to further explore the contact properties of the semiconducting SWCNT and Ti electrode. A temperature dependence on I–V characteristics was observed in the strong hole accumulation region with $V_G = -7.5$ V, as shown in Fig. 4.10a. It was shown that the current I_{DS} decreased with the decrease of the temperature. When the temperature $T > 253$ K, the devices showed ohmic $I - V$ characteristics, with the $I_{DS} - V_{DS}$ curves highly linear and symmetrical. This indicates that the Schottky barriers at the contact are small enough so as to be neglectable for hole transport in the ON-state. When the temperature was decreased, the $I_{DS} - V_{DS}$ curves exhibited nonlinear behavior, which is due to the rectifying action of the Schottky barrier at the low temperature. From the temperature dependence of $I_{DS} - V_{DS}$ curves, we can calculate

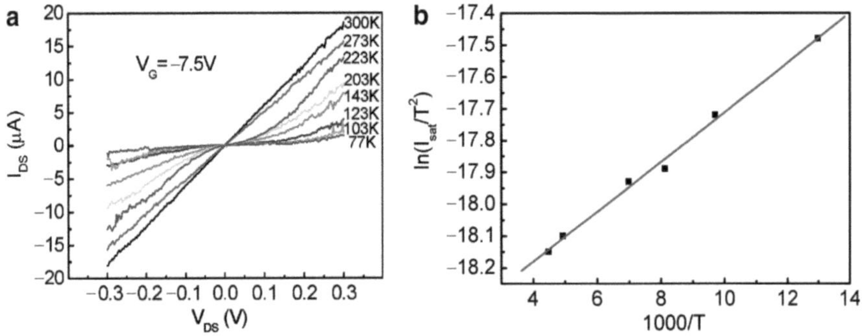

Fig. 4.10 (**a**) I–V Characteristics of a bonded semiconducting SWCNT in a vacuum at different temperatures with $V_G = -7.5$ V. (**b**) Extraction of the Schottky barrier height by a linear fit of the $\ln(I_{sat}/T^2)$ versus 1/T plot. The barrier height is determined by the slope of the linear fit

the Schottky barrier height, ϕ_b, at the contact [24, 30]. With the ideal Schottky barrier junction behavior borrowed, the forward bias current, I_{DS}, as a function of the bias V can be expressed as

$$I_{DS} = I_{sat}[\exp(qV/nkT) - 1], \tag{4.2}$$

where q is the electronic charge, V is the voltage across the junction, I_{sat} is the reverse bias saturation current, and n is known as the ideality factor. I_{sat} is defined as [31]

$$I_{sat} = SAT^2 \exp(-q\phi_b/kT) \tag{4.3}$$

where ϕ_b is the effective barrier height of the junction,S is the diode area, and A is the Richardson constant. Using the (4.3), we constructed a plot of $\ln(I_{sat}/T^2)$ versus $1/T$ (Fig. 4.10b) [15]. From the slope of the plotted data in Fig. 4.10b, the effective barrier height, ϕ_b, is deduced to be about 6.6 meV. This value is much smaller than that (15 meV) obtained by the high-temperature annealing method [24]. The Schottky barrier width, d_b, can be estimated using $d_b = \sqrt{2\varepsilon\varepsilon_0 [(W_{CNT} - W_{Ti} - kT)/e] / eN_A}$ [32], where N_A is the hole concentration that can be estimated from the hole density in the tube, N_h, by using $N_A = 4N_h/\pi d^2$. While N_h can be obtained from $N_h = C_g \cdot V_g^t/eL$ and $C_g = 2\pi\varepsilon_0\varepsilon L/\ln(h/d)$ [33], where C_g is the gate-channel capacitance, V_g^t is the gate voltage required to deplete holes in the tubes, and ε, h, L, and d are the average dielectric constant of the transistor, thickness of silicon dioxide, SWCNT channel length, and the diameter of the SWCNT respectively. For the CNTFET studied here, V_g^t is determined to be ~0.5 V, assuming that holes are sufficiently depleted when the channel conductance decreases six orders of magnitude from the hole accumulation region. Using $W_{Ti} = 4.33$ eV, $W_{CNT} = 4.5$ eV [10], $\varepsilon = 2.5$ [12], $h = 150$ nm, $d = 1.6$ nm, and $L = 1\,\mu$m, it is estimated that $d_b \approx 0.9$ nm at $V_g = 0$. This value is two orders of magnitude smaller than that reported in the literature (~50 nm) [30]. The small barrier height and width indicate that the contact resistance between the semiconducting SWCNT and electrode can be significantly decreased by the ultrasonic bonding treatment.

4.4.4 Mechanical Strength of Nanowelded Contacts

Besides the electrical tests, the mechanical integrity of the welded bonds has also been tested by mechanically pushing the hanging middle segment of a welded nanotube with an AFM tip. In the experiment, 20 unwelded and welded samples were tested. In the contact-mode AFM, with the feedback turned off, the cantilever was pressed down onto the substrate and dragged along the path parallel to the electrode pairs. It was found that when SWCNTs were not welded to the electrodes, SWCNTs were easily shifted and could no longer bridge both electrodes (Fig. 4.11a, b). This observation qualitatively shows that the van der Waals' forces between the

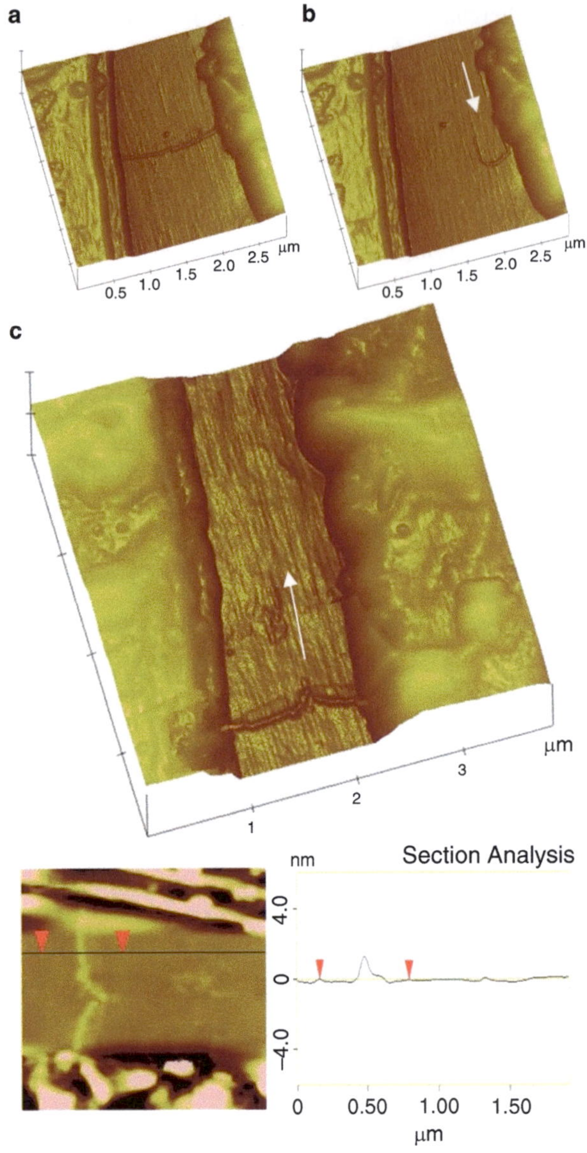

Fig. 4.11 Characterizations of the mechanical stability of the ultrasonic nanowelding by AFM manipulation. (**a**) Three-dimension AFM image of an unwelded SWCNT bridging the electrodes. (**b**) AFM image of the SWCNT in (**a**) shifted laterally by AFM tip. (**c**) AFM image of a nanowelded SWCNT pushed to fracture by AFM tip. The SWCNT diameter was measured to be about 1.5 nm by AFM height, which indicated individual SWCNT instead of SWCNT bundle. The *arrows* in (**b**) and (**c**) show the traveling paths of the AFM tip

CNTs and substrate underneath are not large enough to resist the movement of the CNTs. In contrast, for the nanowelded samples, the two-ends of the 95% SWCNTs remained bonded to the electrodes even after the nanotube had been pushed to fracture (Fig. 4.11c). This suggests that the bonds bear reasonable mechanical strength. It is observed that segments of CNTs close to the electrodes are still approximatively perpendicular to the electrodes while segments of CNTs at the fracture surface are strongly bent after AFM-dragging test, as shown in Fig. 4.11c. This phenomenon can be explained by the existence of the van der Waals force between the CNTs and the substrate [34], which prevents the lateral movement of the CNTs and causes the CNTs to be dragged along its axis, forming this strained configuration [8].

Finally, it should be mentioned that ultrasonic welding experiments have also been performed resulting in the bonding between SiC or Si nanowire and titanium electrode as well as SWCNT and gold electrode. All these bonds show good electrical contacts. Therefore, the present approach is feasible for wide ranges of one-dimensional nanomaterials and metals.

4.5 Conclusion

Two main ways had been adopted to form nanotube–metal contacts in the previous studies of CNT devices. One, the metal electrodes were prepatterned on the substrate first and then the CNTs were deposited onto these metal electrodes; the other is to deposit or grow the CNTs on the substrate first and then find the CNTs and define the metal electrodes on their two ends. The present methods are subject to the problems of the large contact resistance, poor process reliability, or the complex process procedure etc. The simpler, less capital intensive, and more scalable processes are highly desirable.

A novel ultrasonic nanowelding technology has been introduced to reliably bond SWCNTs onto metal electrodes, by pressing SWCNTs against electrodes under a vibrating force at ultrasonic frequency. The bonds formed were measured to be mechanically robust With this technique, the long-term stable low-resistance contact between SWCNTs and electrodes could be achieved and the 2t-resistance was decreased by more than three orders of magnitude, with resistances mainly concentrated around 15 kΩ for a 1-μm-long metallic SWCNT at room temperature. A low effective Schottky barrier height of 6.6 meV between semiconducting SWCNT and Ti electrode in the ON state was obtained after nanowelding. The performance of CNTFETs fabricated by this ultrasonic nanowelding technology had been also significantly improved. A transconductance as high as 3.6 μS for solid-state back-gate individual nanotube FETs could be achieved.

This ultrasonic nanowelding technique does not depend on the specific kind of nanocomponent or metal electrodes. Moreover, it is hopeful that the ultrasonic welding area can be scaled up so that multiple bondings on a substrate can be achieved in one single step, under one single press vibrating at a specific ultrasonic frequency. Therefore, this technology possesses potential for industrial application.

References

1. W. Hoenlein, F. Kreupl, G.S. Duesberg, A.P. Graham, M. Liebau, R.V. Seidel, E. Unger. IEEE. Trans. Comp. Pack. Technol. **27**, 629 (2004).
2. J. Muster, M. Burghard, S. Roth, G.S. Duesberg, E. Hernandez, A. Rubio, J. Vac. Sci. Technol. B **16**, 2796 (1998).
3. M. Burghard, G.S. Duesberg, G. Philipp, J. Muster, S. Roth, Adv. Mater. **10**, 584 (1998).
4. M.F. Yu, O. Lourie, M.J. Dyer, K. Moloni, T.F. Kelly, R.S. Ruoff, Science **287**, 637 (2000).
5. M.F. Yu, M.J. Dyer, G.D. Skidmore, H.W. Rohrs, X.K. Lu, K.D. Ausman, Von., J.R. Her, R.S. Ruoff, Nanotechnology **10**, 244 (1999).
6. P.A. Williams, S.J. Papadakis, M.R. Falvo, A.M. Patel, M. Sinclair, A. Seeger, A. Helser, R.M. Taylor, S. Washburn, R. Superfine, Appl. Phys. Lett. **80**, 2574 (2002).
7. D.N. Madsen, K. Mølhave, R. Mateiu, A.M. Rasmussen, M. Brorson, C.H. Jacobsen, P. Bøggild, Nano Lett. **3**, 47 (2003).
8. C.X. Chen, L.J. Yan, E.S.W. Kong, Y.F. Zhang, Nanotechnology **17**, 2192 (2006).
9. Y. Woo, G.S. Duesberg, S. Roth, Nanotechnology **18**, 095203 (2007).
10. S.J. Tans, M.A.R. Verschueren, C. Dekker Nature (London) **393**, 49 (1998).
11. J.W.G. Wildboer, L.C. Venerma, A.G. Rinzler, R.E. Smalley, C. Dekker, Nature (London) **391**, 59 (1998).
12. M.W. Bockrath, Ph.D. thesis (University of California, Berkeley, 1999).
13. M. Bockrath, D.H. Cobden, P.L. MacEuen, N.G. Chopra, A. Zettl, A. Thess, R.E. Smalley, Science. **275**(5308), 1922 (1997).
14. J. Tersoff, Appl. Phys Lett. **74**(15), 2122 (1999).
15. C.X. Chen L.Y. Liu, Y. Lu, X.J. Sheng, H. Ding, Y.F. Zhang, Carbon **45**, 436 (2007).
16. A. Javey, J. Guo, D.B. Farmer, Q. Wang, D. Wang, R.G. Gordon, Nano Lett. **4**(3), 447 (2004).
17. C.X. Chen, Y.F. Zhang, J. Phys. D: Appl. Phys. **39**, 172 (2006).
18. J.Q. Li, Q. Zhang, D.J. Yang, J.Z. Tian, Carbon, **42**, 2263 (2004).
19. G.G. Harman, *Wire Bonding in Microelectronics: Materials, Processes, Reliability, and Yield*, 2nd edn. (McGraw-Hill, New York, 1997).
20. S.Y. Kang, P.M. Williams, T.S. McLaren, Y.C. Lee, Mater. Chem. Phys. **42**, 31 (1995).
21. Ph. Avouris, Chem. Phys. **281**, 429 (2002).
22. C.D. Breach, F. Wulff, Microelectron. Reliab. **44**, 973 (2004).
23. F.W. Wulff, C.D. Breach, D. Stephan, Saraswati, K.J. Dittmer, Proceedings of 6th Electronics Packaging Technology Conference, Singapore 2004, p. 348.
24. R. Martel, V. Derycke, C. Lavoie, J. Appenzeller, K.K. Chan, J. Tersoff, Ph. Avouris, Phys. Rev. Lett. **87**, 256805 (2001).
25. A. Bachtold, P. Hadley, T. Nakanishi, C. Dekker, Science **294**, 1317 (2001).
26. J. Appenzeller, R. Martel, V. Derycke, M. Radosavljevic, S. Wind, D. Neumayer, Ph. Avouris, Micro. Eng. **64**, 391 (2002).
27. V. Derycke, R. Martel, J. Appenzeller, Ph. Avouris, Appl. Phys. Lett. **80**, 2773 (2002).
28. S.J. Wind, J. Appenzeller, R. Martel, V. Derycke, Ph. Avouris, Appl. Phys. Lett. **80**, 3817 (2002).
29. A. Javey, Q. Wang, A. Ural, Y. Li, H. Dai Nano Lett. **2**, 929 (2002).
30. K. Xiao, Y. Liu, P. Hu, G. Yu, Y. Sun, D. Zhu J. Am. Chem. Soc. **127**, 8614 (2005).
31. B.L. Sharma, *Metal–Semiconductor Schottky Barrier Junctions and Their Applications* (Plenum Press, New York, 1984) p. 60.
32. S.M. Sze, *Physics of Semiconductor Devices* (Wiley, New York, 1981) p. 248.
33. A. Bezryadin, A.R.M. Verschueren, S.J. Tans C. Dekker, Phys. Rev. Lett. **80**, 4036 (1998).
34. H.C. Postma, A. Sellmeijer, C. Dekker, Adv. Mater. **12**, 1299 (2000).

Chapter 5
Nanowelded Multichannel Carbon-Nanotube Field-Effect Transistors (MC-CNTFETs)

5.1 Introduction

The silicon-based electronic technology has made quite a great progress in the past five decades since the invention of the integrated circuit [1]. The advance is maintained primarily through the size scaling of the devices, i.e., metal–oxide–semiconductor field-effect transistor (MOSFET), which has resulted in many successive generations of devices with increased transistor performance and density. In case of the continual decrease for the device dimensions at a present speed, most of scientists forecast that the development of the integrated circuit will meet its physical and theoretical limit soon, perhaps in next decade [2–4]. To keep the device performance continuing to update, the seeking and preparation of new technologies is mandatory [5].

So far, two distinct routes have been taken to address that issue. One of them is to adopt the revolutionary technologies based on totally new concepts, e.g., two-terminal molecular devices [6], quantum computing [7], spintronics [8], and so on. However, these technologies will be incompatible with present application developed from silicon-based electronic industry. In the other case, attention is paid to a more evolutionary approach that is based on the well-established three terminal transistor concept, but utilizes alternative materials, specially single-walled carbon nanotubes (SWCNTs) that possess many unique advantages including one-dimensional nanoscale structure as well as the excellent electrical and optical properties [5]. High-mobility, low-defect structure and intrinsic nanometer scale of carbon nanotubes (CNTs) have led to an intense research effort into the viability of utilizing carbon-nanotube field-effect transistors (CNTFETs) as a replacement for, or a complement to, future semiconductor devices [9–17].

CNTFETs prepared have shown their great superiority in performance vis-a-vis state-of-the-art silicon devices. Nevertheless, before the practical implementation of CNTFETs in CMOS circuit application, many key issues remain to be solved such as the single nanotube channeled device structure and the large metal–nanotube contact resistance, which limit further upgrade of the device performance. Superior performance for the CNTFETs can be expected if multiple parallel CNTs with good electrical contacts and spatial control can be utilized to construct the CNTFETs. This multichanneled device structure can satisfy the occasions of applications where the high transconductance or large current output is required e.g., power transistors or sensors. CNTFETs fabricated with an individual SWCNT usually exhibit small current outputs and low transconductances due to the limited maximum current carrying capacity (~25 μA) of individual SWCNT [9]. Although the SWCNT bundle contains numerous SWCNTs, the FETs prepared by an individual SWCNT bundle cannot still effectively improve this problem, because it is not only that the SWCNTs in the core of the bundle cannot be well electrically contacted by the electrodes but also that the screen effect of charges exists between adjacent nanotubes [10]. Theoretically, FETs constructed by a periodic array of carbon nanotubes will exhibit a much higher performance than those constructed by individual SWCNT [10, 11]. Another merit of the multichanneled structure is that it can greatly enhance the applicability and reliability of the fabricated devices and circuits. Some big defects might exist on some SWCNTs, which will cause the devices fabricated with these individual SWCNTs to be easy to fail and decrease the applicability and reliability of the devices. Using the multiple SWCNTs as the conduction channel will effectively avoid this problem, because if only one of channels in the device works the device will be able to keep operating normally. In addition, by varying the channel number in the CNTFETs, the drain current and transconductance of the device can be adjusted to meet the different requirements in the circuit, which will play a role analogous to changing the gate width of traditional Si MOSFETs.

Although there have been reports of high output current CNTFETs that use a large number of CNTs as the channel recently [18, 19], their general device performance was not good enough due to the lack of techniques to produce well directed, dispersed CNT channels and good metal–nanotube contacts. Furthermore, the realization of reliable, ohmic contacts between a large-scale SWCNT array and electrodes also presents a new challenge, due to the fact that many previously reported methods were basically of small-scale spot-treatment nature, having rigorous processing condition [20]. These technical choke points prevent the achievement of high-performance CNT-based devices. In this chapter, the multichanneled CNTFET (MC-CNTFET) is introduced and discussed. In the MC-CNTFET, the dispersive, parallel SWCNT array with a controllable density, which is formed by the AC electric field-assisted alignment, is used as the channel to construct the device. Compared with the single-channel CNTFET (SC-CNTFET), the prepared MC-CNTFET not only exhibits the larger current output and transconductance, but also possesses higher reliability and applicability. When the SWCNT channels are nanowelded onto the source and drain electrodes by the ultrasonic nanowelding technique, the device performance of the MC-CNTFETs can be effectively enhanced. Using the MC-CNTFETs, complementary logic inverters with a high gain are also built.

5.2 Present Optimization Methods for CNTFETs

Since the first CNTFET was fabricated in 1998 (see Fig. 5.1) [21], CNTFETs have experienced great advances at the device structure as well as device performance [22]. Various methods had been attempted to optimize the devices. These methods

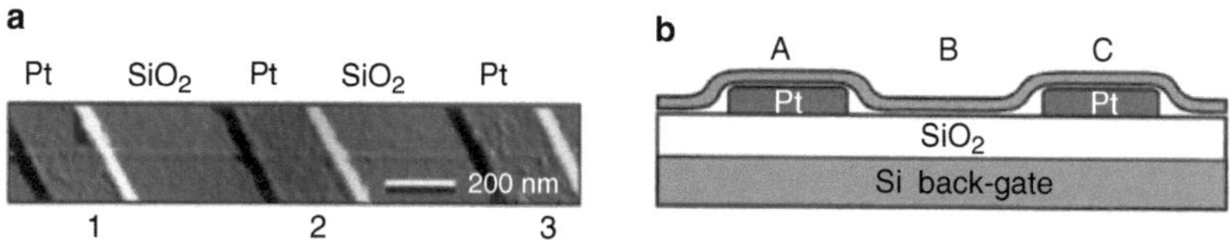

Fig. 5.1 (**a**) Tapping-mode atomic force microscopy (AFM) image of an individual carbon nanotube on top of three Pt electrodes. (**b**) Schematic side view of the TUBEFET device. Images adapted from [21]

can mainly be divided into four aspects: (1) Decrease the contact resistance between CNTs and metal electrodes; (2) Increase the tuning efficiency of gate to CNT channels; (3) Shorten the CNT channel length; (4) Adopt the optimized device structure. Representative researches and updated progress on the optimization methods of CNTFETs are introduced in this section. From the part, the advances in CNTFETs can be also learned in the rough.

5.2.1 *Decrease the Contact Resistance Between SWCNTs and Metal Electrodes*

When SWCNTs are placed on the metal electrodes, SWCNTs are contacted to electrodes by the weak van der Waals force, which causes a large contact resistance between SWCNTs and electrodes so as to degrade the device performance.

Many ways had been attempted to decrease the contact resistance. Martel et al. adopted a high-temperature anneal method to improve the contact [23]. They fabricated the metal electrodes on two ends of SWCNTs and then annealed the devices at a high temperature (~800°C), which makes the SWCNTs and metal electrodes form the compound. With the high-temperature treatment, the contact resistance could be decreased to $\sim 30\,\mathrm{k\Omega}$, the device current increased from the nanoampere level to the microampere level, and the transconductance ($g_m \sim 0.34\,\mu\mathrm{S}$) was enhanced 200 times. Zhou et al. grew the individual semiconducting SWCNT between source and drain electrodes by CVD method and then fabricated the metal electrodes on SWCNTs [24]. The fabricated CNTFETs exhibited a transconductance of 200 nS and a gain of 3. In addition, the low-resistance contacts can also be obtained by choosing the suitable metal as contact electrodes. Au is found to be able to form reliable and good contacts with SWCNTs [25]. And Pd and semiconducting SWCNTs can form the Ohmic contact, which decreases the contact resistance significantly [26].

5.2.2 *Increase the Tuning Efficiency of Gate to SWCNT Channels*

Another important method to improve the performance of CNTFETs is increasing the tuning efficiency of gate voltage to SWCNT channels. At present, there are mainly three ways to enhance the gate tuning efficiency.

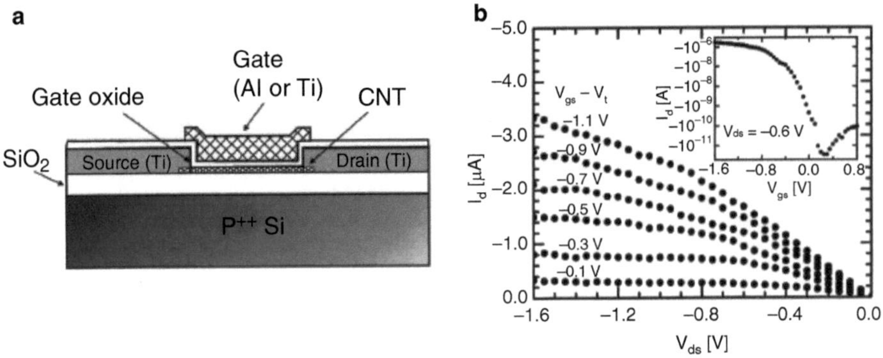

Fig. 5.2 (**a**) Schematic cross section of top gate CNTFET showing the gate and source and drain electrodes. (**b**) Output characteristic of a top gate p-type CNFET with a Ti gate and a gate oxide thickness of 15 nm. The gate voltage values range from −0.1 to −1.1 V above the threshold voltage, which is −0.5 V. *Inset*: Transfer characteristic of the CNTFET for $V_{ds} = -0.6$ V. Images adapted from [27]

5.2.2.1 Use the Top-Gate Structure

For previous back-gate CNTFETs, a thick gate insulator (100–150 nm) was used, which generally causes a large threshold voltage. Moreover, the SWCNTs in the back-gate devices were exposed to air and the gate capacitance was diluted by the air with the low dielectric constant, which causes poor tuning efficiency for gate bias. In addition, each CNTFETs on the chip cannot be controlled separately for the back-gate structure. To solve the disadvantages, Wind et al. developed a top-gate CNTFET (Fig. 5.2) [27]. At first, they fabricated the source and drain electrodes on two ends of SWCNT. Then, a gate dielectric was deposited from a mixture of SiH_4 and O_2 by CVD method on the SWCNT. Finally, a Ti or Al electrode was overlaid on the SWCNT channel. It was demonstrated that the top-gate structure effectively enhances the tuning effect of gate bias. The threshold voltage of the device (−0.5 V) is far lower than that (−12 V) of back-gate device, and the drive current and transconductance ($g_m \sim 0.34\,\mu$S) were also increased effectively.

5.2.2.2 Decrease the Thickness of Gate Insulator

Decreasing the distance of gate electrode and SWCNT can also increase the gate capacitance and thereby enhance the tuning effect of gate electrode. Although the principle for this way is relatively simple, the process is not easy to realize. Bachtold et al. successfully fabricated the CNTFET with a few nanometer-thick gate insulator [28]. They utilized a naturally oxided Al_2O_3 layer on the Al electrode as the gate

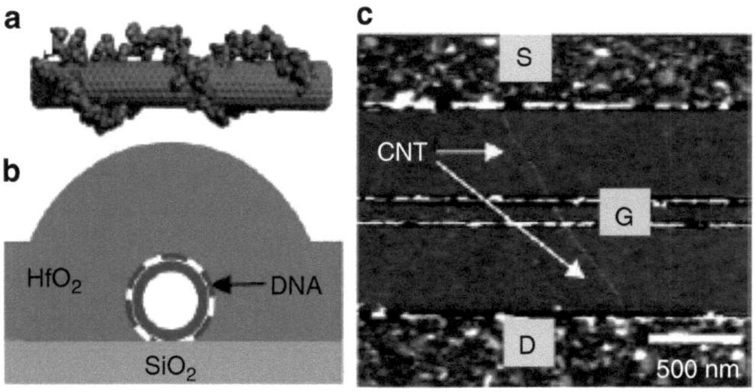

Fig. 5.3 The CNTFETs fabricated by depositing the HfO_2 on DNA-functionalized SWCNTs with atomic layer deposition (ALD) technique. (**a**) Schematic of a DNA coated SWCNT. (**b**) Cross-sectional view of HfO_2 ($\sim$3 nm by ALD) conformally deposited on a DNA functionalized nanotube lying on a SiO_2 substrate. (**c**) AFM image of a high k SWCNT FET with top-gate (G) underlapping source (S) and drain (D). Images adapted from [29]

insulator, the thickness of which is estimated to be a few nanometers. The fabricated CNTFETs exhibited a transconductance of 0.3 μS, an On-state current of $\sim$100 nA, a gain of >10, and a current On/Off ratio of 10^5. Lu et al. had fabricated the top-gate CNTFETs with the ultrathin high k dielectrics by depositing the HfO_2 on DNA-functionalized SWCNTs with atomic layer deposition (ALD) technique (Fig. 5.3) [29]. With the DNA functionalization, the high-performance nanotube-high k FETs free of gate-leakage currents were reliably obtained with HfO_2 thickness down to 2–3 nm. The fabricated CNTFETs could reliably achieve the subthreshold slope $S = 60$ mV per decade at room temperature.

5.2.2.3 Adopt the High-*k* Gate Insulator

For the traditional MOSFETs, the gate capacitance C_G, and gate insulator thickness t_{ox}, have a relation of $C_G \propto 1/t_{ox}$. Different from the MOSFET, the gate capacitance of the CNTFET can be expressed as $C_G \approx 2\pi\varepsilon\varepsilon_0/\ln(2h/r)$, where ε, h, and r are the dielectric constant, thickness of silicon dioxide, and the radius of the carbon nanotube respectively. This slowly logarithmic dependence between C_G and h for the CNTFETs degrades the effect of reducing the gate insulator thickness. Moreover, the too thin gate insulator will cause the unexpected large leakage current. Therefore, ones begin appealing to using high-k gate dielectric to increase the gate capacitance.

Appenzeller et al. used HfO_2 ($k \sim 11$) to displace SiO_2 as the gate insulator to fabricate the back-gate CNTFETs [30]. The 20-nm-thick HfO_2 layer was deposited on the n-type Si substrate by CVD method under 400°C. The fabricated CNTFETs exhibit a drive current of 270 μA μm^{-1}, a maximum transconductance of 425 μS μm^{-1} and an inverse subthreshold slope of 140 mV dec^{-1}. Javey et al. utilized 8-nm-thick high-k ZrO_2 ($k \sim 25$) as the gate insulator to fabricate the

CNTFETs. In the experiment, an individual SWCNT prepared by CVD method was connected between two 3-μm-gap Mo source and drain electrodes, which were patterned on Si substrate thermally coated by 500-nm-thick SiO_2 layer. And then the thin ZrO_2 was deposited on the SWCNT with the atom layer deposition (ALD) technology. Finally, the 2-μm-width Ti/Au top electrode was fabricated on the ZrO_2 gate insulator. The fabricated CNTFETs exhibited high device performances: For the p-type CNTFETs, the inverse subthreshold slope of $S \sim 70\,\mathrm{mV\,dec^{-1}}$ was obtained, which approximated the theory limitation ($\sim$60 mV) of field-effect transistors under room temperature; the transconductrance was 12 μS for individual nanotube transistors ($3{,}000\,\mathrm{S\,m^{-1}}$ normalized by double nanotube diameter); The carrier mobility of the devices reached $3{,}000\,\mathrm{cm^2\,V^{-1}\,s^{-1}}$, which is more than eight times the p-type Si bulk mobility $\left(450\,\mathrm{cm^2\,V^{-1}\,s^{-1}}\right)$. For n-type CNTFETs, an inverse subthreshold slope of $S \sim 90\,\mathrm{mV\,dec^{-1}}$ was achieved. The gain of the constructed CMOS with this CNTFETs reached 60, which is the highest value reported so far.

On this basis, Javey et al. integrated the Ohmic contacts and high-k gate dielectric technology into the fabrication of CNTFETs to further improve the device performance [26]. In the device, the 8-nm-thick high-k HfO_2 ($k \sim 20$) dielectric layer was used as the top gate insulator, which was deposited by ALD technique and Pd metal was adopted to contact the semiconducting SWCNTs. Figure 5.4 shows the schematic diagram of this CNTFET. In this structure of CNTFETs, the heavily doped Si substrate served as back gate to electrostatically dope the SWCNT. The two ends of doped SWCNT acted as the source and drain electrodes and the SWCNT segment under the top gate served as the conduction channel. The fabricated device exhibited a transconductance of $g_m \sim 20\,\mu\mathrm{S}$ ($5{,}000\,\mathrm{S\,m^{-1}}$ normalized by double nanotube diameter), an On-state saturation current of $I_{ON\_sat} \sim 15\,\mu\mathrm{A}$ $\left(\sim 3750\,\mu\mathrm{A}\,\mu\mathrm{m^{-1}}\right)$, and an On-state linear conductance

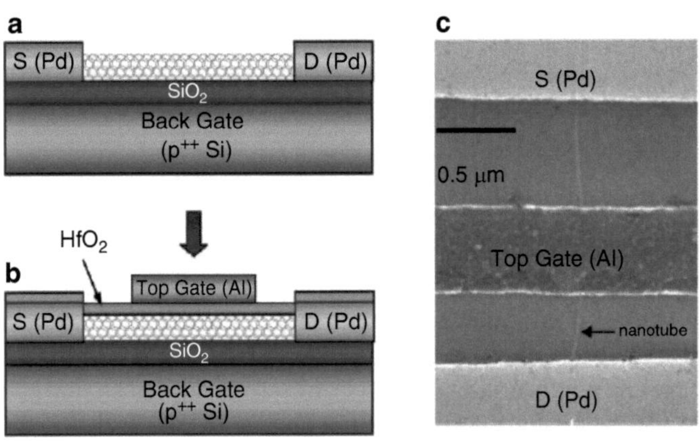

Fig. 5.4 The CNTFET fabricated by integrating the Ohmic contact and high-k gate insulator. (**a**) Schematic device drawings for a nanotube contacted by metal Pd. (**b**) Nanotube FET with HfO_2 fabricated by atomic layer deposition (ALD) as top gate insulator. (**c**) Scanning electron microscopy (SEM) image of the device depicted in (**b**). Images adapted from [26]

of $G_{ON} \sim 0.1 \times 4e^2/h$. A rough estimation shows that under the similar gate tuning g_m and I_{ON_sat} are five times higher than that of Si p-MOSFET and three times higher than that of the same structure of CNTFET with Mo as contact metal. An inverse subthreshold slope of S $\sim 80\,\text{mV dec}^{-1}$ and a current ratio of $I_{ON}/I_{MIN} > 10^4$ were achieved by the device. Moreover, the bipolar behavior was also effectively restricted in this CNTFET.

5.2.3 Shorten the SWCNT Channel Length

Another way to enhance the CNTFET performance is shortening the SWCNT channel length. When the SWCNT channel is long (more than several micrometers), the carriers will be scattered in the transport process, which makes the carrier mobility reduced [31, 32]. Previous studies had shown that the free path of carriers for the SWCNT could be up to about 500 nm. Therefore, if CNT channel is decreased to a length of less than this value, the carriers will conduct a ballistic transport in the SWCNT. Thus, the mobility of the device will be greatly increased and the CNTFET fabricated will be effectively enhanced. For this reason, many short channel CNTFETs had been fabricated and researched. Seidel et al. reported a CNTFET with an 18-nm-length channel [33]. They used HSQ (hydrogensilsesquioxane) as the mask to successfully fabricate the very small channel gap (a little more than 10 nm). Under a drain bias of 0.4 V, the fabricated CNTFET exhibited an On-state current of >15 μA, a transconductance of 6.75 μS, and a current ratio of $>10^6$.

However, it should be noted that the improvement effect of this method is degraded when the length of SWCNT channels is smaller than the free path of carriers, because the carriers transport ballistically in the SWCNTs and the device mobility is determined mainly by the contact property for this time. Previous studies [34, 35] had shown that when the channel length is smaller than ~260 nm, the influence of the lateral scaling on the performance of CNTFETs is little.

5.2.4 Adopt the Optimized Device Structure

The novel device structures are also adopted to enhance the performance of CNTFETs. Lin et al. [36] designed a double-gate CNTFET. Figure 5.5 shows the schematic diagram and subthreshold characteristics of this CNTFET. In this CNTFET, the switch and the injected carrier type were controlled by the Al gate and the Si back gate respectively. As the effect of Si gate was screened by the Al electrode, SWCNT in the B region would be regulated only by the Al gate, which caused an ideal switch behavior. An inverse subthreshold slope of $\sim 63\,\text{mV dec}^{-1}$ approximating the theory value and a low OFF current of <100 fA can be achieved.

Especially, a coaxial surrounding-gate CNTFET (see Fig. 5.6a) is predicted to have the excellent device performance, in which the CNT is enwrapped by a thin

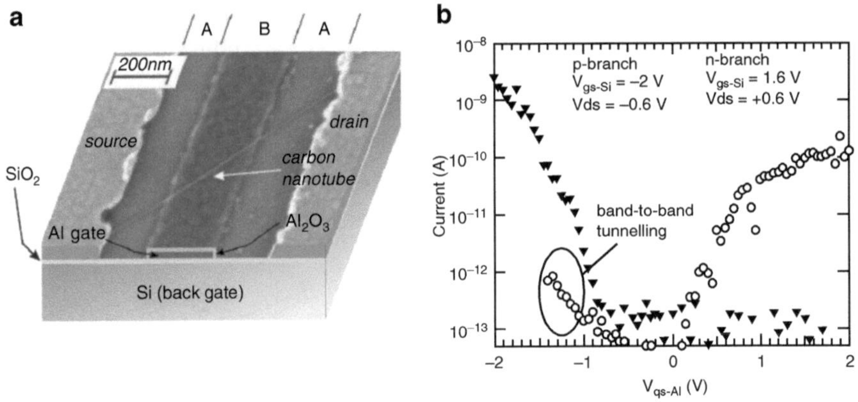

Fig. 5.5 (**a**) Composite of the device layout of a dual-gate CNFET, showing the SEM image of a CNFET with an Al middle gate underneath the nanotube. (**b**) Subthreshold characteristics (I_d–V_{gs-Al}) of the dual-gate CNFET measured at constant Si gate voltages $V_{gs-Si} = -2\,V$ (*filled down triangle*) and $+1.6\,V$ (*open circle*), exhibiting clear p- and n-type unipolar behaviors, respectively. Images adapted from [36]

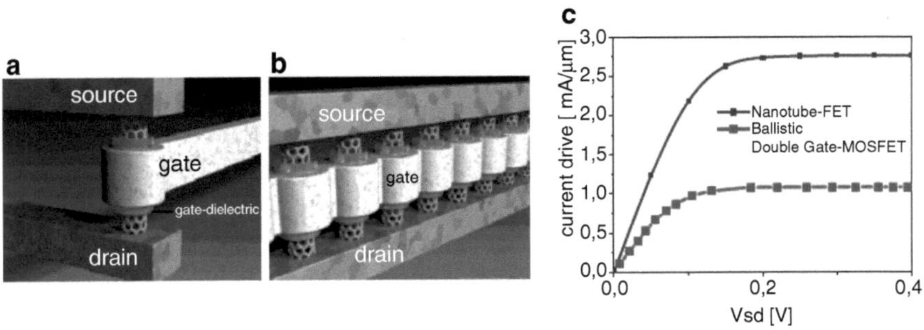

Fig. 5.6 (**a**) Vertical CNTFET concept, (**b**) parallel operation of many VCNTFETs, (**c**) show output characteristic of VCNTFET is superior to the ballistic double gate Si-MOSFET. Images adapted from [37, 38]

insulator layer with an equivalent thickness of ~1 nm. The device performance is greatly enhanced because of the efficient gate tuning effect for this CNTFET. This coaxial surrounding-gate CNTFET will be hopeful to be used to construct the vertical integrated structure (see Fig. 5.6b) [37]. Compared with the state-of-the-art best ballistic double-gate Si-MOSFET, this vertical CNTFET shows the superior device performance (Fig. 5.6c), which is enough to meet the requirement of the ITRS roadmap in 2016 for traditional Si-MOSFETs [38].

In summary, with these optimization methods, a great research progress for CNTFETs has been gained. Although some difficulties such as the separation of semiconducting and metallic CNTs still need to be broken through before the industrial application of CNTFETs, the experimental and theoretic study results have shown its bright prospect.

5.3 NonNanowelded MC-CNTFETs

In order to improve the device performance of CNTFETs, a MC-CNTFET mentioned in Sect. 5.1 has been developed by our research group. In the MC-CNTFET, an array of parallel SWCNTs is used to act as the conduction channels of CNTFETs. This MC-CNTFET not only can meet the requirement of large output current and high transconductance, but also can manifest good reliability and applicability. It is also found that the transconductance of the MC-CNTFET has an almost linear dependency on the SWCNT channel number, which opens up a promising way to tune the transconductance of FETs by controlling the channel number.

To fabricate multichanneled field-effect transistors, well-dispersed and aligned SWCNT array is needed. Seidel et al. [33] fabricated the planar FET consisting of large amount of SWCNTs by CVD method, which allowed ON currents in the order of several milliamperes. However, due to the lack of good separation and alignment, the metallic SWCNTs were difficult to be burned off due to the shielding function of the surrounding semiconducting SWCNTs and the obtained FETs exhibited a low On/Off current ratio of about 500. Besides, because of the entanglement of SWCNTs, the semiconducting SWCNTs were easily destroyed or burned off by local overheating in the process of burning the metallic SWCNTs, which caused a tradeoff between maximum On-current and On/Off ratio. Recently, the electrical-field-assisted alignment has been demonstrated as an effective method for the orientated deposition of CNTs [39–42]. With this technology, CNTs dispersed in a suspension can be deposited directly between the electrodes by applying the DC or AC electric field to both face-to-face electrodes after the CNT suspension is introduced onto the patterned wafer. Nevertheless, in the previous research, the CNT suspension are obtained by directly ultrasonicating CNTs in various solvents, which is adverse to the dispersivity of CNTs and thus the quality of the directed alignment is unsatisfactory. In this section, dispersed, aligned carbon nanotube array is formed between electrodes by directed placement of surface decorated SWCNTs with the AC dielectrophoresis method. The multichanneled field-effect transistors are fabricated with these dispersed SWCNTs, which act as the channels of the FET. The fabricated MC-CNTFETs were shown to have a performance superior to that of the SC-CNTFETs [43].

5.3.1 Device Fabrication

5.3.1.1 Surface Decoration of SWCNTS

Purified SWCNTs with the average diameter of 1.4 nm were dispersed in H_2SO_4–HNO_3 (3:1) and kept at 80°C for 30 min in the reflux system to introduce the carboxyl at their opening ends and the defect locations on their walls. Following that, a certain volume of octadecylamine (ODA) and N,N'-dicyclohexylcarbodiimide (DCC) was mixed with the SWCNTs and reactd at 90°C

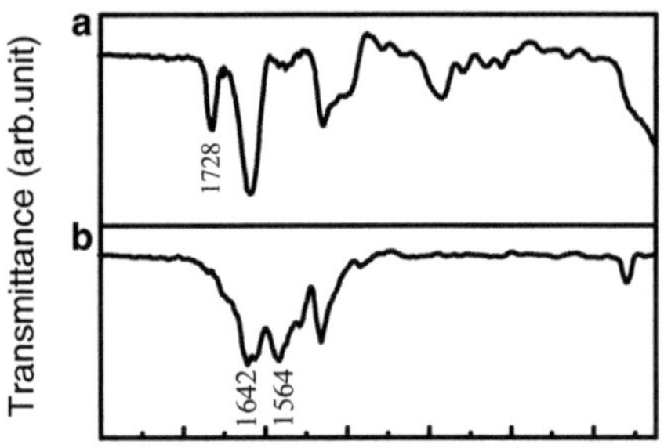

Fig. 5.7 Infrared spectrograms of the treated SWCNTs in different stages. (**a**) After mixed acid treatment and (**b**) after decoration

for 24 h, which made the ODA grafted onto SWCNTs by condensation reaction. Afterwards, the excess ODA and DCC was removed by rinsing with ethanol for five times (10 min sonication at 40 kHz). After filtration, the product was dried in vacuum at room temperature.

Figure 5.7 shows the infrared absorption spectrogram for the SWCNTs at different treatment stages. The stretching vibration of C=O bond in carboxyl $\left(1{,}728\,\mathrm{cm}^{-1}\right)$ is observed after the mixed acid treatment (see Fig. 5.7a), which shows that the carboxyl is introduced onto SWCNTs. After reacted with ODA and DCC, the stretching vibration mode of C=O bond in carboxyl disappears, while the peak at $1{,}642\,\mathrm{cm}^{-1}$ assigned to the stretching vibration of C=O bond in amide appears. Besides, the peaks at $1{,}564\,\mathrm{cm}^{-1}$ corresponding to the in-plane bending vibration of N–H bond are also observed in the spectrum (Fig. 5.7b). These results indicate that the SWCNTs are grafted by ODA. The mole percent of grafted ODA to C in SWCNT can be expressed as $\delta = \frac{(m_2 - m_1)/M_{\mathrm{ODA}}}{m_1/M_{\mathrm{C}}} \times 100\%$, where m_1 and m_2 are respectively the mass of SWCNTs before and after the predecoration and M_{ODA} and M_{C} are respectively the mole mass of ODA and C. The mass of SWCNTs was measured to be respectively 31.0 and 45.7 mg before and after the predecoration at 90°C. Hence, the mole percent, δ, of the grafted ODA was about 2.13%.

The decorated SWCNTs exhibited substantial high solubility in organic solvents and could be easily ultrasonically dissolved in chloroform or isopropanol to form a stable SWCNT solution. The SWCNT solutions were well dispersed and no precipitation was observed with the concentration up to $300\,\mu\mathrm{g\,ml}^{-1}$ upon prolonged standing, which attributes to the later directed placement. In addition, the surface decoration may contribute to weaken the mutual attractive force between SWCNTs, causing exfoliation of the SWCNT bundles to give individual nanotubes [44].

5.3.1.2 Alignment of SWCNTs Between Electrodes

Source and drain electrodes were made of 100-nm-thick Au on n-type silicon wafers with a 500-nm thermally oxidized layer. The gap between source and drain of the CNTFETs was 1 μm and the gate width was 10 μm. After applying a 10 V AC bias with a frequency of 5 MHz between source and drain, a droplet of SWCNT solution

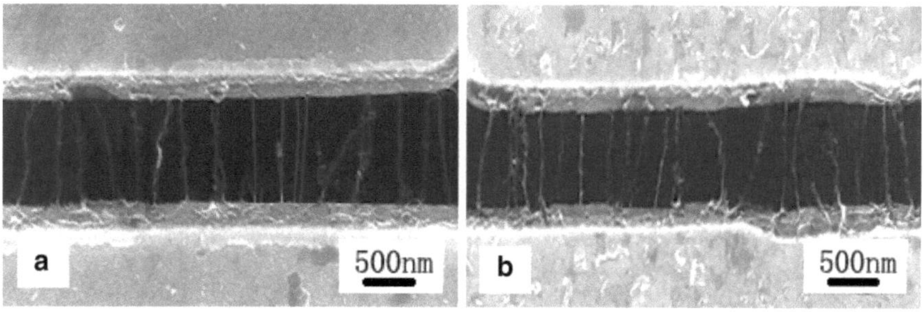

Fig. 5.8 SEM images of the SWCNT channels aligned respectively in high-frequency AC electric field by using solutions of chloroform (**a**) and isopropanol (**b**), respectively

(typical SWCNTs concentration $= 0.5\,\mu g\,ml^{-1}$) was introduced onto the structured wafer. The AC bias was not switched off until the solution was evaporated up and remained aligned SWCNTs between the electrodes on the wafer. Finally, the SWCNTs were irradiated with UV light with 194 nm wavelength for 60 min to remove the ODA on the SWCNTs [45, 46]. After the alkyl chains were decomposed and evaporated up from the substrate, a good contact between SWCNTs and electrodes was formed.

Figure 5.8 shows the scanning electron microscope (SEM) image of the SWCNT channels that were aligned in chloroform (Fig. 5.2a) and isopropanol (Fig. 5.2b), respectively. It is observed that the SWCNTs are approximately perpendicular to electrodes and the average spaces between neighboring SWCNTs are about 200 and 150 nm respectively. In contrast to these results, when the SWCNTs of the same concentration ultrasonicated in the chloroform without the predecoration are aligned with AC electric field, the SWCNTs are entangled although visible orientation is also observed. Therefore, the surface decoration contributes to obviating the entanglement of SWCNTs and form the dispersed aligned SWCNT array. To account for this, aside from the reason for the greatly improved solubility and dispersibility of SWCNTs in the solution by the predecoration, it is presumed that some net charges may be held by the grafted SWCNTs after the predecoration. It is the repulsion between these net charges on SWCNTs that causes SWCNTs to align dispersedly and periodically.

In the alignment processes, two forces effect the movement of SWCNTs. One is an electrophoretic force coming from the interaction of the electric field and nonzero net charges [47] and the other is the dielectrophoretic force produced by the interaction of a nonuniform electric field with the electric field-induced dipole. In an AC electric field, the movement due to the electrophoresis force is oscillatory with zero time-average and only dielectrophoretic force causes the movement of SWCNTs. The dielectrophoretic force drives SWCNTs to move toward regions of high electric field along the electric field gradient. The time-averaged dielectrophoretic force can be expressed as $\left\langle \vec{F}_{\mathrm{DEP}} \right\rangle = 2\pi\varepsilon_{\mathrm{m}} a^3 \,\mathrm{Re}\,[\tilde{\alpha}\,(\omega)]\, \nabla \,|E_{\mathrm{rms}}|^2$, where ε_{m} is the dielectric constant of the solvent medium, a is the particle radius, and $\mathrm{Re}\,[\tilde{\alpha}\,(\omega)]$ is the real part of the complex, Clausius–Mossotti factor (i.e., $(\tilde{\varepsilon}_p - \tilde{\varepsilon}_m)/(\tilde{\varepsilon}_p + 2\tilde{\varepsilon}_m)$, where ε_p is the permittivity of the particle) [48]. Thus, SWCNTs are deposited around the electrodes where the electric field strength is highest by the translational movement. Besides, SWCNTs also experience a rotary movement during the alignment process, because the induced dipoles also bear a torque force at the same time. Thus, SWCNTs are ultimately aligned in the direction of applied electric field by adjusting themselves along tube axis parallel to the electric field direction, as shown in the SEM images.

In the experiment, the DC electric field was also attempted to align the modified SWCNTs. It was found that when a DC bias was used to align SWCNTs, although SWCNTs could also be directly aligned between the electrodes, some of them were only contacted to the positively biased electrode and could not make the reliable electrical connection with both electrodes. This could be due to that some negative charges were held on SWCNTs after surface modification, which interacted with the DC electric field and caused SWCNTs to be attracted toward the positively biased electrode. More contaminations were also observed to deposit around the electrodes when using DC electric field than AC electric field. This result was consistent to the earlier reports that high-frequency AC electric field has a function of purifying the raw SWCNTs. It was found that the existence of the contamination would make the breakdown of the metallic SWCNTs more difficult in the following process of device fabrication, and therefore was disadvantageous to the achievement of good I–V characteristics. Therefore, it is preferable to use the high-frequency AC electric field to directly align the SWCNTs.

5.3.1.3 Breakdown of Metallic SWCNTs

As the aligned SWCNTs were a mixture of metal and semiconductor, an electrical breakdown method [49, 50] was used to burned off the metallic SWCNTs. In this experiment, a 15 V bias was applied first on the back gate to completely deplete the carriers of semiconducting SWCNTs, and then a sufficiently large current was applied to burn off the undepleted metallic SWCNTs. FETs with multiple channels of SWCNTs were finally obtained.

5.3.2 Device Characteristics

5.3.2.1 The Performance Compare of MC-CNTFETs and SC-CNTFETs

Figure 5.9a, b shows typical I–V characteristics of as-prepared MC-CNTFET by aligning 24 SWCNTs between the source and the drain. The drain current, I_{DS}, decreased with the increase of V_G when the drain voltage, V_{DS}, was constant and the CNTFETs was switched off when the $V_G > 1$ V, which indicates a p-type transistor. The device exhibited an ON drain current of $I_{ON} \sim 59.0\,\mu$A at

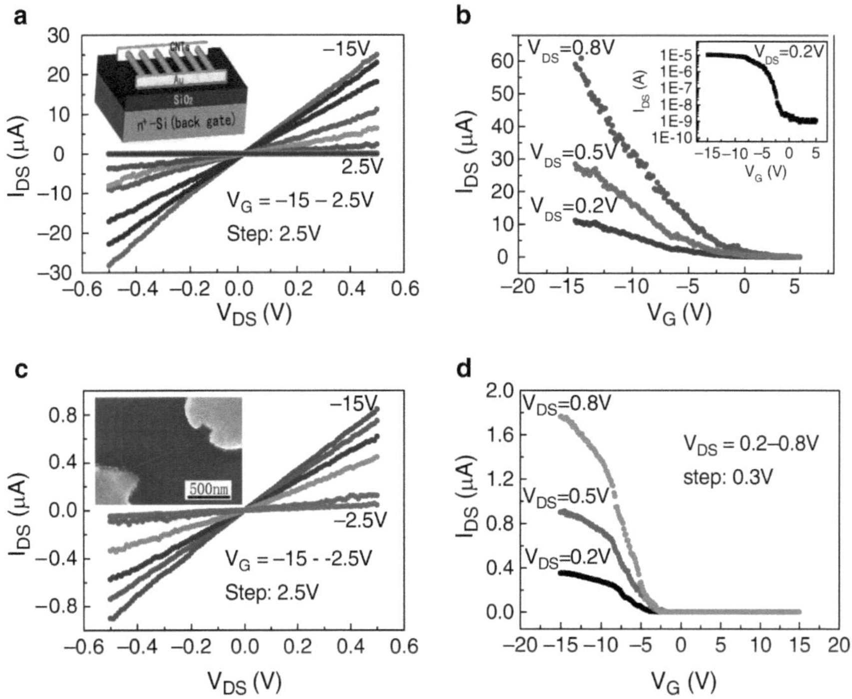

Fig. 5.9 *I–V* characteristics of the MC-CNTFET and the SC-CNTFET. (**a**) I_{DS}–V_{DS} curves of the MC-CNTFET. *Inset*: Structural schematic diagram. (**b**) I_{DS}–V_G curves of the MC-CNTFET. *Inset*: Log (I_{DS}) vs. V_G at $V_{DS} = 0.2$ V. (**c**) I_{DS}–V_{DS} curves of the SC-CNTFET. *Inset*: SEM image of an individual SWCNT bridging the electrodes. (**d**) I_{DS}–V_G curves of the SC-CNTFET

$V_G = -15$ V when $V_{DS} = 0.8$ V. The transconductance dI_{ds}/dV_g in linear region of $-15\,\text{V} < V_G < -8$ V is about 5.43 μS for $V_{DS} = 0.8$ V. An On–Off ratio, higher than 10^4, is obtained (as shown in the inset of Fig. 5.9b).

A comparison is done between the MC-CNTFET and the SC-CNTFET. Figures 5.9c, d display the output and transfer characteristics of the SC-CNTFET prepared by the same fabrication process with the further diluted SWCNT solution. It is indicated that the ON current and transconductance are greatly lower for the SC-CNTFET. The transconductance of the SC-CNTFET in linear region of $-7.5\,\text{V} < V_G < -4$ V is only 0.248 μS at $V_{DS} = 0.8$ V and decreases about 22 times. A larger threshold voltage is shown for the multichanneled FET, perhaps because the inconsistent turn-off performance for different SWCNT channels in the multichanneled FET, which requires a relatively larger gate voltage to deplete the carriers in all SWCNT channel. The I_{DS} of the SC-CNTFET tends to saturate for $V_G < -7.5$ V, while I_{DS} saturation does not occur until V_G was decreased to -15 V for the MC-CNTFET.

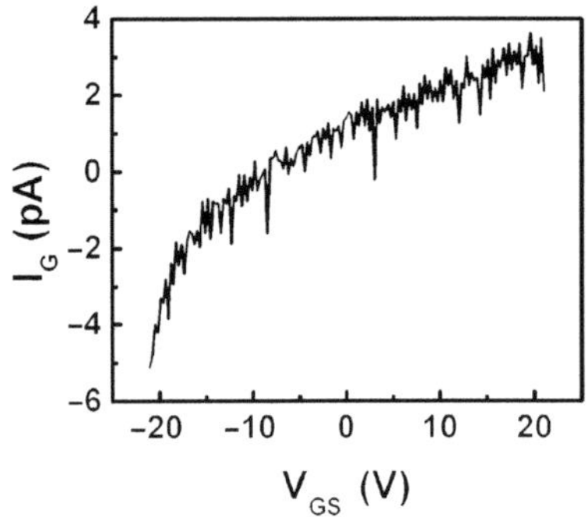

Fig. 5.10 Gate leak current versus back-gate voltage for the fabricated multichanneled FET

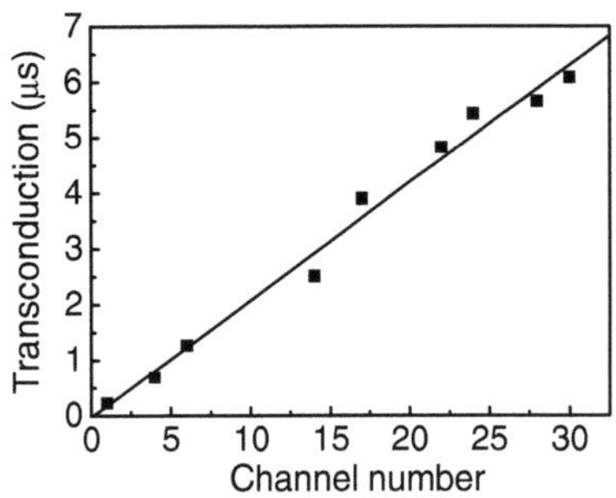

Fig. 5.11 The relation of the transconductance and the SWCNT channel number. Data were taken at $V_{DS} = 0.8$ V

The fabricated multichanneled FET has a low gate leak current (at the picoampere level) when the gate voltage is lower than 20 V (Fig. 5.10), which is far smaller than I_{DS} at the same V_G values so as to be negligible.

5.3.2.2 The Influence of the Channel Number

In order to understand the dependence between transconductance and channel numbers, a serial of the SWCNT MC-CNTFETs with the different channel numbers is fabricated. It is found that the transconductance rises almost linearly with the channel number increasing (Fig. 5.11). Further studies are needed to elucidate it and here a preliminary explanation is provided. For an individual SWCNT on the planar insulator, the gate-channel capacitance is $C_g = 2\pi L\varepsilon\varepsilon_0/\ln(2h/r)$, where ε, h, and r are the dielectric constant thickness of silicon dioxide, and the radius of the carbon nanotube respectively. In our FETs, SWCNTs dispersedly bridge the source and drain electrodes with a definite space between the neighboring SWCNTs. According to [52], when the space between neighboring nanotubes is larger than double diameter of the nanotubes, the interaction between SWCNTs can be ignored and the gate-channel capacitance for the parallel SWCNTs is almost proportional to the SWCNT number. Therefore, the gate-channel capacitance of the multichanneled structure should be $C_g^{/} = nC_g$, where n is the channel number. On the other hand, in the linear $I_{DS} - V_{DS}$ region, the transconductance of the CNTFET can be expressed as $\mathrm{d}I_{\mathrm{ds}}/\mathrm{d}V_{\mathrm{g}} = \mu_{\mathrm{h}} C_{\mathrm{g}}^{/} V_{\mathrm{ds}}/L^2$, where μ_{h} is the mobility rate of charge carrier and L is the length of the CNT between the source and drain electrodes [51]. Thus, it can be deduced that the transconductance is proportional to the SWCNT number, n, for the fixed L and V_{DS}.

This result opens up a promising way to tune the transconductance of CNTFETs linearly by controlling the SWCNT channel number, which is similar to changing the gate-width of the conventional Si FETs. By this way, the transconductance can be adjusted to meet different requirements in an integrated circuit.

5.3.3 Device Reliability

It is also noticed that the multichanneled structure can improve the reliability the FETs and enhance the yield of the device. In the experiment, 20 MC-CNTFETs and 20 SC-CNTFETs are measured in the ambient atmosphere at intervals of a definite period. It is shown that the SC-CNTFETs are generally out of work after a term of about three weeks, while the MC-CNTFETs can keep a stable performance after several months. The reason for this is that the device still can keep operating when some channels break down in a MC-CNTFET, while the device will fail if the only channel breaks down in a SC-CNTFET. In the case of only considering the influence of the channel, the yield of the MC-CNTFET can be given by $y = \sum_{l=0}^{n-m} C_n^l k^l (1-k)^{n-l}$, where k is the failure rate of per individual SWCNT channel caused by fatal factors and n and m are respectively the total fabricated channel number and the needed successful channel number for meeting the requirement of normal operation. It is indicated that the yield of the device is sharply increased by the SWCNT channel number.

5.4 Nanowelded MC-CNTFETs

In order to further improve the device performance, the ultrasonic nanowelding technology had been used to optimize the device in our previous work. In this section, we will look at the MC-CNTFET with nanowelded contacts.

5.4.1 Device Fabrication Processes

SWCNTs with an average diameter of 1.6 nm were synthesized by an arc-discharge process, followed by a purification and sufficient dispersion in 1,2-dichloroethane. The source and drain electrodes with the parallel finger arrays were fabricated with standard UV lithography and lift-off technique, as depicted in Fig. 5.6a. The electrodes were made of Cr/Au (15/135 nm) or Al (150 nm) on n-type silicon wafers with a 150-nm thermally oxidized layer. The finger pairs in the electrodes are 0.7-μm wide, 5-μm separate, and have a gap of 1 μm. An AC dielectrophoresis method was used to position the SWCNTs onto metal electrodes [53]. A drop of the SWCNT suspension was introduced onto the structured wafers, after a high-frequency AC bias with a peak-to-peak voltage of 10 V and a frequency of 5 MHz was applied to the source and drain electrodes. The wafers were heated to 60°C in order to enhance the evaporation speed of the 1,2-dichloroethane. The AC bias was switched off after the 1,2-dichloroethane solvent was evaporated. Then, an ultrasonic nanowelding technique was applied to bond the SWCNT channels and metal electrodes [54]. An Al_2O_3 single crystal with a 150-μm-diameter pressing surface and a rms surface roughness of 0.2 nm was used as the welding head. A force of 55.2 mN was applied by the welding head to press the SWCNTs against electrodes. At the same time, an ultrasonic with a frequency of 60 kHz and a power of 0.15 W was applied to the welding head through an ultrasonic transducer. The process sustained a duration of 0.2 S. Under the combined effect of the ultrasonic and press, the SWCNTs were embedded and welded into the metal electrodes.

Figure 5.12b shows the images of SWCNTs bridging source and drain electrodes. The ends of the nanowelded SWCNTs were embedded into the electrodes and as such, SWCNTs were almost invisible on the electrodes, as shown in Fig. 5.12c. The formation of the nanowelding between SWCNTs and the electrodes can be attributed to the "acoustic softening effect" of the ultrasonic to the metal [54], which has been explained in Sect. 4.3.2. The high-frequency ultrasonic energy can greatly enhance the activity of metal atoms and soften the metal so as to cause plastic deformation of the metal under the clamping stress, which makes the one-dimension nanosized SWCNTs "embeddable" and weldable into the metal electrodes.

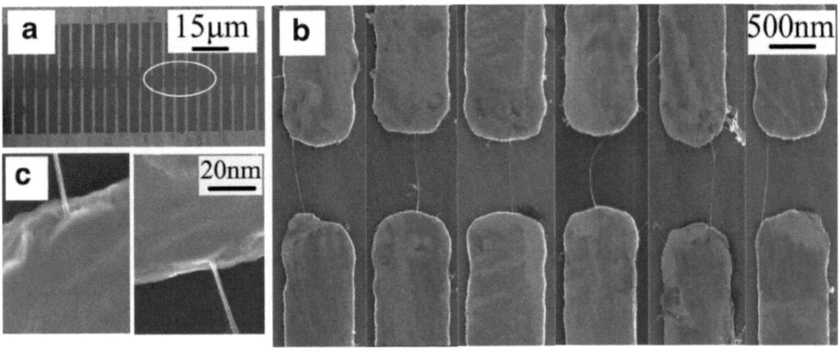

Fig. 5.12 (**a**) Low-magnification SEM image of an electrode array with 22 diametrically opposed finger pairs. (**b**) Zoom-in view of a local region at the electrode gap, showing that each finger pair is bridged by one SWCNT. (**c**) SEM images of the SWCNT ends nanowelded onto the metal electrodes

---

The research found that for the electrodes with 22 finger pairs, 20 of them could be bridged by the SWCNTs, giving a yield of about 91%. Most of electrode pairs were bridged by 1 SWCNT. Only 2 out of them are contacted by 2 SWCNTs. Therefore, this method enables us to control the number and space of SWCNT channels by simply changing the number and space of the finger pairs. The formation of this separate and aligned array can be attributed to that SWCNTs are driven by dielectrophoretic force to move toward higher electric field area along the direction of the electric field strength gradient [53], and thereby ultimately deposited between the finger pairs where the electric field strength is the highest.

After selectively destroying the metallic SWCNTs in the array by an electrical breakdown method [55], the as-prepared MC-CNTFETs with the degenerately doped silicon substrate as the gate were tested.

5.4.2 Nanowelded p-type MC-CNTFETs

The typical $I-V$ characteristics of MC-CNTFETs are shown in Fig. 5.13. It is indicated that the MC-CNTFETs with nanowelded contacts exhibit excellent general device performance, greatly superior to the nonnanowelded SC-CNTFET prepared in the same manner (Table 5.1). The MC-CNTFETs contacted by Au electrodes showed p-type characteristics. As shown in the inset of Fig. 5.13, I_{DS}–V_{DS} curves are highly linear and symmetrical for $-0.3\,V < V_{DS} < 0.3\,V$ in the ON state. For an Au-contacted MC-CNTFET with 15 SWCNT channels, note that at a low drain bias of $V_{DS} = -0.15\,V$, the high ON current of $I_{ON} \sim 235\,\mu A$(16 μA for average per channel when normalized by channel number) and transconductance of $g_m \sim 50.2\,\mu S$ (3.3 μS average per channel) are achieved. This normalized transconductance greatly exceeds those previously reported for solid-state back-gate individual nanotube FETs [9, 11, 51, 56–58]. The ON resistance obtained is as low as $0.64\,k\Omega$, decreased by more than four orders of magnitude than nonnanowelding

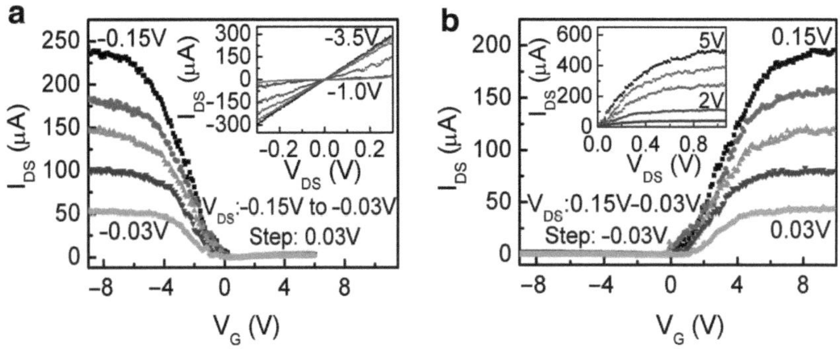

Fig. 5.13 (**a**) Transfer characteristic of the Au-contacted MC-CNTFET. *Inset*: Output characteristic measured with $V_G = -3.5$ to $-1.0\,V$ in steps of 0.5 V. (**b**) Transfer characteristic of the Al-contacted MC-CNTFET. *Inset*: Output characteristic measured with $V_G = 5$ to 0 V in steps of $-1\,V$

Table 5.1 Key device parameters of the MC- and SC-CNTFETs

	Nonnanowelded Au-contacted	Nonnanowelded Au-contacted	Nonnanowelded Au-contacted
Channel length (μm)	1	1	1
Channel number	1	15	13
Drain bias (V)	−0.15	−0.15	0.15
I_{ON} (μA)	0.014	235	196
g_m (μS)	0.028	50.2	36.5
R_{ON} (kΩ)	10.714	0.64	0.77
Mobility[a] ($cm^2\,V^{-1}\,s^{-1}$)	64	7.160	5.311
I_{ON}/I_{OFF}	10^5	10^5–10^6	10^6–10^7
S (mV dec^{-1})	529	116	92
Threshold voltage (V)	−2.5	−0.55	0.65

[a]The field-effect mobility μ is derived at the drain bias of $V_{DS} = \pm 10\,mV$. A gate-channel capacitor of ~0.502fF and ~0.436fF is deduced respectively for the p-MC-CNTFET with 15 channels and n-MC-CNTFET with 13 channels

SC-CNTFETs. A field-effect hole mobility, which is independent of the channel number, has been deduced to reach as high as $\mu_h \sim 7,160\,cm^2\,V^{-1}\,s^{-1}$. This mobility is the among the best for SWCNT-FETs presently reported [10, 32, 56], which is about 68 times higher than that of Au-contacted CNTFETs without optimizing contacts ($\sim 106\,cm^2\,V^{-1}\,s^{-1}$) [51]. This indicates that the contact resistance is significantly decreased and high transparent transport has been achieved.

The devices exhibited an inverse subthreshold slope of $S \sim 116\,mV\,dec^{-1}$. Although this value is a little inferior to recently reported values (63–70 $mV\,dec^{-1}$) [12, 59] due to that the top-gate structure and high-κ gate insulator are used in those literatures, it is far superior to the back-gated SC-CNTFETs with the contacts improved by other method ($S \sim 730\,mV\,dec^{-1}$) [57]. A high on–off current ratio of 10^5–10^6 has also been achieved (the inset in Fig. 5.13).

The reason for the significant decrease of the contact resistance by the ultrasonic nanowelding is given in Sect. 4.4.1. The high transconductance and good switch performance for our devices can be attributed to the improvement of contact characteristics. Besides the formation of the reliable low-resistance contacts, the transistor configuration is also changed after nanowelding [54]. Before nanowelding, the SWCNTs are "side-bonded" to the electrodes. While after nanowelding, the "end-bonded" nanotube-metal contact is formed. The latter can induce a stronger electric field at the junction and effectively enhance the modulation effect of the gate [58].

5.4.3 *Nanowelded n-Type MC-CNTFETs*

Behaving rather differently from the Au-contacted MC-CNTFETs, the Al-contacted MC-CNTFETs show n-type characteristics (Fig. 5.13), which agrees with the report in [42]. The n-MC-CNTFETs with 13 SWCNT channels exhibit a large

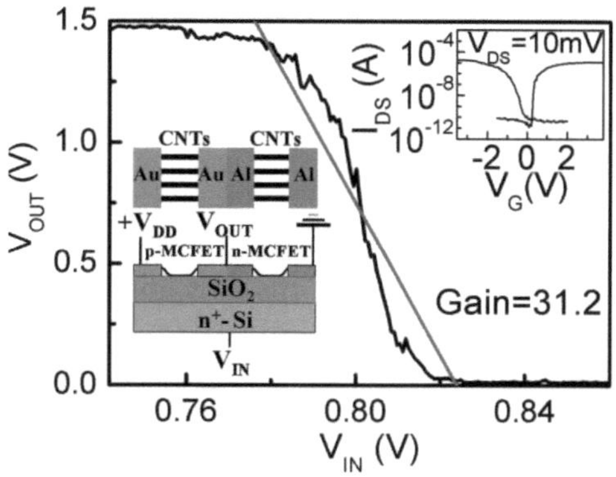

Fig. 5.14 Transfer characteristic of a complementary voltage inverter constructed by the p- and n-MC-CNTFETs corresponding to Figure 5.2. The *red line* shows an output–input gain of 31.2. *Left inset*: The schematic diagram of the complementary MC-CNTFET inverter. *Right inset*: Transfer characteristics measured with $V_{DS} = -10$ and $+10\,\text{mV}$ for the p- and n-MC-CNTFETs, respectively

$I_{ON} \sim 196\,\mu\text{A}$ ($\sim 15\,\mu\text{A}$ per channel), a high $g_m \sim 36.5\,\mu\text{S}$ ($\sim 2.8\,\mu\text{S}$ per channel), and a low $S \sim 92$ (see Table 5.1). Note that a high $I_{ON}/I_{OFF} \sim 10^6$–10^7 was achieved for the transistor with no significant ambipolar behavior (the inset in Fig. 5.3). An electron mobility of up to $\sim$5,311 $\text{cm}^2\text{V}^{-1}\,\text{s}^{-1}$ has been reached, much higher than the previously reported Al-contacted CNTFET ($\sim$3,750 $\text{cm}^2\text{V}^{-1}\,\text{s}^{-1}$) [56]. These characteristics are excellent compared with previous n-type back-gate CNTFETs [11, 56, 60, 61].

5.4.4 SWCNT Complementary Logic Inverters

Using our technique, the complementary CNTFET inverters can be readily constructed by fabricating the p- and n-MC-CNTFETs respectively on the connected Au and Al electrode pairs via the lithography overlay. Figure 5.14 shows the transfer characteristic of a complementary voltage inverter constructed by the aforementioned p- and n-MC-CNTFETs. It is indicated that an ideal voltage switch characteristic has been achieved. The low- and high-level outputs were very close to 0 V and V_{DD} (1.5 V) respectively. A large voltage gain of up to 31.2 was achieved, which significantly exceeds those recently reported for complementary individual-SWCNT inverters with local bottom gates ($\sim$8) [62] and complementary inverters constructed by SWCNT thin film transistors ($\sim$8) [13]. The achievement of the large gain can be attributed to the high performance of the p- and n-MC-CNTFETs, which have the high values of transconductance.

5.5 Conclusion

The optimization methods and present research advance of CNTFETs have been reviewed. These optimization methods can be mainly summarized as follows: (1) Decreasing the contact resistance of CNTs and source and drain electrodes; (2) Increasing the tuning effect of gate to CNT channels; (3) Shortening the CNT channel length; and (4) Optimizing the device structure.

Individual SWCNTs had been utilized as the conduction channels to construct CNTFETs. However, the single-channel device structure would cause many shortcomings such as limited output current and transconductance, poor device reliability and applicability, and incapable scaling for the gate width. In order to solve these key problems in the fabrication of CNTFETs, a MC-CNTFET with dispersed, directed SWCNTs as the conduction channels has been developed. The fabrication processes, device characteristics, and effects of the CNT channel number are discussed in the chapter. The device performance of the MC-CNTFET is also compared with that of the SC-CNTFET. The obtained MC-CNTFETs not only exhibit the larger output current and transconductance, but also hold higher reliability and applicability. The transconductance of the MC-CNTFET is found to have a linearly proportional dependency on the SWCNT channel number, which promises a potential application in the integrated circuits.

The contacts between CNTs and metal electrodes also play a key role in the device performance of fabricated CNTFETs. To improve the device performance, the ultrasonic nanowelding technology is applied to optimize the device. Both high-performance p- and n-MC-CNTFETs have been successfully built with the use of the ultrasonic nanowelding technique, which can form the firm contacts between SWCNTs and metal electrodes and significantly improve the nanotube–metal contact characteristics. The complementary logic inverter constructed by the nanowelded MC-CNTFETs has been demonstrated to exhibit a high gain up to 31.2.

References

1. http://en.wikipedia.org/wiki/Integrated_circuit#cite_note-3
2. H.S. Momose, M. Ono, T. Yoshitomi, T. Ohguro, S. Nakamura, M. Saito, in *IEDM Tech.*, San Francisco, CA, 1994, p. 593.
3. B. Doris, in *IEDM Tech Dig[C]*, San Francisco, 2002, p. 267.
4. H. Iwai, Solid State Electron. **48**, 497 (2004).
5. S.V. Rotkin, S. Subramoney, *Applied Physics of Carbon Nanotubes: Fundamentals of Theory, Optics and Transport Devices* (Springer, New York, 2005).
6. N.M. Kriplania, D.P. Nackashi, C.J. Amsinck, N.H.D. Spigna, M.B. Steer, P.D. Franzon, R.L. Rick, G.C. Solomon, J.R. Reimers, Chem. Phys. **326**(1), 188 (2006).
7. C.H. Bennett, D.P. Divincenzo, Nature **404**, 267 (2000).
8. D. Sarma, J. Fabian, X. Hu, I. Zutic, IEEE Trans. Magn. **36**, 2821 (2000).
9. A. Bathtold, P. Hadley, T. Nakanishi, C. Dekker, Science **294**, 1317 (2001).
10. A. Javey, H. Kim, M. Brink, Q. Wang, A. Ural, J. Guo, P. McIntyre, P. McEuen, M. Lundstrom, H. Dai, Nat. Mat. **1**, 241 (2002).

11. V. Derycke, R. Martel, J. Appenzeller, P.H. Avouris, Nano Lett. **1**, 453 (2001).
12. Y.M. Lin, J. Appenzeller, J. Knoch, Ph. Avouris, IEEE Trans. Nanotechnol. **4**, 481 (2005).
13. S.H. Hur, M.H. Yoon, A. Gaur, M. Shim, A. Facchetti, T.J. Marks, J.A. Rogers, J. Am. Chem. Soc. **127**, 13808 (2005).
14. H. Oh, J.J. Kim, W. Song, S. Moon, N. Kim, J. Kim, N. Park, Appl. Phys. Lett. **88**, 103 (2006).
15. M.H. Yang, K.B.K. Teo, L. Gangloff, W.I. Milne, D.G. Hasko, Y. Robert, P. Legagneux, Appl. Phys. Lett. **88**, 113 (2006).
16. Z.H. Chen, D. Farmer, S. Xu, R. Gordon, P. Avouris, J. Appenzeller, IEEE Electron. Dev. Lett. **29**, 183 (2008).
17. A. Raychowdhury, A. Keshavarzi, J. Kurtin, V. De, K. Roy, IEEE Trans. Electron. Dev. **53**, 2711 (2006).
18. R. Seidel, A.P. Graham, E. Unger, G.S. Duesberg, M. Liebau, W. Steinhoegl, F. Kreupl, W. Hoenlein, W. Pompe, Nano Lett. **4**, 831 (2004).
19. E.S. Snow, P.M. Campbell, M.G. Ancona, J.P. Novak, Appl. Phys. Lett. **86**, 33105 (2005).
20. Q. Chen, S. Wang, L.M. Peng, Nanotechnology **17**, 1087 (2006).
21. S.J. Tans, A.R.M. Verschueren, C. Dekker, Nature **393**, 49 (1998).
22. C.X. Chen, Y.F. Zhang, Open Nanosci. J. **1**, 13 (2007).
23. R. Martel, H.S.P. Wong, K. Chan, P. Avouris, IEDM Tech. Dig. 159 (2001).
24. C.W. Zhou, J. Kong, H. Dai, Appl. Phys. Lett. **76**, 1597 (2000).
25. F. Nihey, H. Hongo, Y. Ochiai, M. Yudasaka, S. Iijima, Jpn. J. Appl. Phys. **42**, L1288 (2003).
26. A. Javey, J. Guo, D.B. Farmer, Q. Wang, D. Wang, R.G. Gordon, M. Lundstrom, H. Dai, Nano Lett. **4**, 447 (2004).
27. S.J. Wind, J. Appenzeller, R. Martel, V. Derycke, P. Avouris, Appl. Phys. Lett. **80**, 3817 (2002).
28. A. Bathtold, P. Hadley, T. Nakanishi, C. Dekker, Science **294**, 1317 (2001).
29. Y.R. Lu, S. Bangsaruntip, X. Wang, L. Zhang, Y. Nishi, H. Dai, J. Am. Chem. Soc. **128**, 3518 (2006).
30. J. Appenzeller, R. Martel, V. Derycke, M. Radosavjevic, S. Wind, D. Neumayer, P. Avouris, Micro. Eng. **64**, 391 (2002).
31. J. Appenzeller, J. Knoch, M. Radosavljevic, P. Avouris, Phys. Rev. Lett. **92**, 226802 (2004).
32. M.S. Fuhrer, B.M. Kim, T. Durkop, T. Brintlinger, Nano Lett. **2**, 755 (2002).
33. R.V. Seidel, A.P. Graham, J.R.B. Kretz, B. Rajasekharan, G.S. Duesberg, M. Liebau, E. Unger, F. Kreupl, W. Hoenlein, Nano Lett. **5**, 147 (2005).
34. S.J. Wind, J. Appenzeller, P. Avouris, Phys. Rev. Lett. **91**, 058301 (2003).
35. S.J. Wind, M. Radosavljevic, J. Appenzeller, P. Avouris, J. Vac. Sci. Technol. B **21**, 2856 (2003).
36. Y.M. Lin, J. Appenzeller, J. Knoch, Z. Chen, P. Avouris, IEEE Trans. Nanotechnol. **4**, 481 (2005).
37. W. Hoenlein, F. Kreupl, G.S. Duesberg, A.P. Graham, M. Liebau, R.V. Seidel, E. Unger, IEEE Trans. Comp. Pack. Technol. **27**, 629 (2004).
38. J. Guo, M. Lundstrom, S. Datta, Appl. Phys. Lett. **80**, 3192 (2002).
39. X.Q. Chen, T. Saito, H. Yamada, K. Matsushige, Appl. Phys. Lett. **78**, 3714 (2001).
40. K. Yamamoto, S. Akita, Y. Nakayama, J. Phys. D: Appl. Phys. **31**, L34 (1998).
41. R. Krupke, F. Hennrich, H. Lohneysen, M. Kappes, Science **301**, 344 (2003).
42. L. Nagahara, I. Amlani, J. Lewenstein, R.K. Tsui, Appl. Phys. Lett. **80**, 3826 (2002).
43. C.X. Chen, Y.F. Zhang, J. Nanosci. Nanotechnol. **6**, 3789 (2006).
44. J. Chen, M.A. Hamon, H. Hu, Y.S. Chen, A.M. Rao, P.C. Eklund, R.C. Haddon, Science **282**, 95 (1998).
45. Y.Z. Guo, M. Mitsuishi, T. Miyashita, Macromolecules **34**, 3548 (2001).
46. Y.Z. Guo, J. Wu, Y.F. Zhang, Chem. Phys. Lett. **362**, 314 (2002).
47. H.H. Gommans, J.W. Alldredge, H. Tashiro, J. Park, J. Magnuson, A.G. Rinzler, J. Appl. Phys. **88**, 2508 (2000).
48. H. Morgan, N.G. Green, in *AC Electrokinetics: Colloids and Nanoparticles -Microtechnologies and Microsystems Series 2* (PA Research Studies, Philadelphia, USA, 2003) p. 50.
49. P.G. Collins, M.S. Arnold, Ph. Avouris, Science **292**, 706 (2001).
50. P.G. Collions, M. Hersam, M. Arnold, R. Martel, Ph. Avouris, Phys. Rev. Lett. **86**, 3128 (2001).

51. J.Q. Li, Q. Zhang, D.J. Yang, J.Z. Tian, Carbon **42**, 2263 (2004).
52. J. Guo, S. Goasguen, M. Lundstrom, S. Datta, Appl. Phys. Lett. **81**, 1486 (2002).
53. C.X. Chen, Y.F. Zhang, J. Phys. D: Appl. Phys. **39**, 172 (2006).
54. C.X. Chen, L.J. Yan, E.S.W. Kong, Y.F. Zhang, Nanotechnology. **17**, 2192 (2006).
55. P.G. Collins, M.S. Arnold, P.H. Avouris, Science **292**, 706 (2001).
56. A. Javey, Q. Wang, W. Kim, H. Dai, Proc. IEDM 741 (2003).
57. R. Martel, H.S.P. Wong, K. Chan, P.H. Avouris, Proc. IEDM 159 (2001).
58. R. Martel, V. Derycke, C. Lavoie, J. Appenzeller, K. Chan, J. Tersoff, Ph. Avouris, Phys. Rev. Lett. **87**, 256805 (2001).
59. M.H. Yang, K.B.K. Teo, L. Gangloff, W.I. Milne, D.G. Hasko, Y. Robert, P. Legagneux, Appl. Phys. Lett. **88**, 113 (2006).
60. H. Oh, J.J. Kim, W. Song, S. Moon, N. Kim, J. Kim, N. Park, Appl. Phys. Lett. **88**, 103 (2006).
61. Y. Nosho, Y. Ohno, S. Kishimoto, T. Mizutani, Appl. Phys. Lett. **86**, 73105 (2005).
62. A. Javey, Q. Wang, A. Ural, Y. Li, H. Dai, Nano Lett. **2,** 929 (2002).

Chapter 6
Carbon-Nanotube Solar Photovoltaic Microcells with Nanowelded Contacts

6.1 Introduction

Great research interests have been shown on photovoltaic (PV) cells over the past four decades or so for their vast application in the domains of energy and communication [1–3]. To achieve efficient, low-cost PV cells, the selection of the ideal PV material, especially energy conversion material, and the optimized device structure are critical. In numerous materials, the semiconducting single-walled carbon nanotubes (SWCNTs) are attractive as the energy conversion material for PV applications due to their unique structure and excellent photoelectric properties. Almost defect-free structure and strong one-dimensional quantum local confinement effect of SWCNTs can decrease the recombination probability of photogenerated carriers and prolong its relaxation time [4, 5]. As a result, the main limitation factors of the efficiency of the traditional silicon PV device can be overcome. A wide band-gap distribution due to variation in diameter enables SWCNTs to match the whole solar spectrum and enhance the absorption of solar energy [6]. Furthermore, all semiconducting SWCNTs have direct band gaps, which make the photoexcitation process easily conducted without needing the assistance of phonons [7].

SWCNT films had been attempted to fabricate the photoelectric chemical solar cells in previous literature [8]. However, because of the lack of the efficient separation and collection of photoexcited carriers and large intertube interaction, the maximum monochromatic incident photo-to-current conversion efficiency (IPCE) acquired for the cell is only 0.15%. In this chapter, we introduce a novel SWCNT PV solar microcells. In this cell, a directed array of monolayer SWCNTs was nanowelded onto two asymmetrical metal electrodes with high and low work function, causing a strong built-in electric field in SWCNTs for efficiently separating photogenerated electron–hole pairs. The monolayer SWCNT PV cell shows a power conversion efficiency (η) of ~0.80% and 0.31% at the solar-light illumination of 8.8 W cm^{-2} and 100 mW cm^{-2} respectively. Correspondingly, a high internal η of 12.6% and 5.1% was estimated for SWCNTs in the device based on the simulation of the actual absorbed incident power by SWCNTs. A promising application potential for SWCNTs in PV devices can be shown with the results.

6.2 Traditional Application of Carbon Nanotubes in Solar Photovoltaic Cells

6.2.1 Using Carbon Nanotubes as Transport Path

In previous studies, carbon nantobues (CNTs) had been employed in the donor–acceptor (D–A) type photovoltaic cells to improve the device performance [9–12]. In this cell, the conjugated polymers such as poly(3-octylthiophene) (P3OT) or poly (3-hexylthiophene) (P3HT) blended with CNTs are used to fabricate the devices. The conjugated polymers act as the photoactive material and CNTs act as the acceptor of the dissociated electrons or holes and their transport path. The device architecture of a photovoltaic cell based on SWCNT-conjugated polymer, poly(3-octylthiophene) composites [9] is shown in Fig. 6.1. The selection of CNTs in the device can be attributed to its extremely high surface area $\left(\sim 1{,}600\,\mathrm{m^2\,g^{-1}}\right)$ [13], which offers a tremendous opportunity for exciton dissociation, and its easy formation of percolation pathways in the polymer at a low doping level, which provides the means for high carrier mobility and efficient charge transfer. When CNTs are mixed with the polymer, the internal junctions between the polymer (electron donating and hole accepting) and the electron acceptor molecule (hole donating) are created. This allows the preferential transfer of the electrons into the electron acceptor molecule while leaving the holes to be preferentially transported through the polymer: a process known as photoinduced charge transfer. Figure 6.2 gives the I–V characteristics of an ITO/P3OT–SWNTs/Al photovoltaic device obtained in the dark and under illumination by Kymakis et al. The nanotube–polymer blend device with a low nanotube concentration (<1%) shows an open-circuit voltage (V_{oc}) of 0.75 V. The short-circuit photocurrent density (I_{sc}) of the device is about $0.12\,\mathrm{mA\,cm^{-2}}$, which increases by two orders of magnitude compared with the device prepared by the pristine polymer, and the fill factor (FF) also increases from 0.3 to 0.4. The power efficiency of the blend device is dramatically increased from 2.5×10^{-5} to 0.04% with respect to the pristine one. The enhancement in the photovoltaic properties of the blend device is due to the introduction of internal polymer–nanotube junctions within the polymer matrix. These junctions act as dissociation centers,

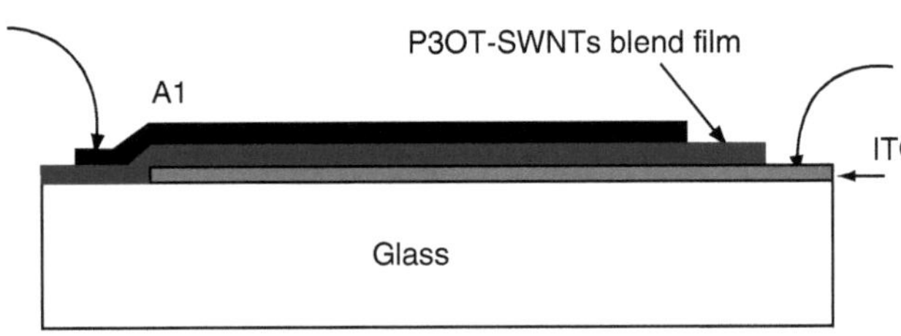

Fig. 6.1 The device architecture of the ITO/P3OT-SWNTs/Al photovoltaic cell. Adapted from [9]

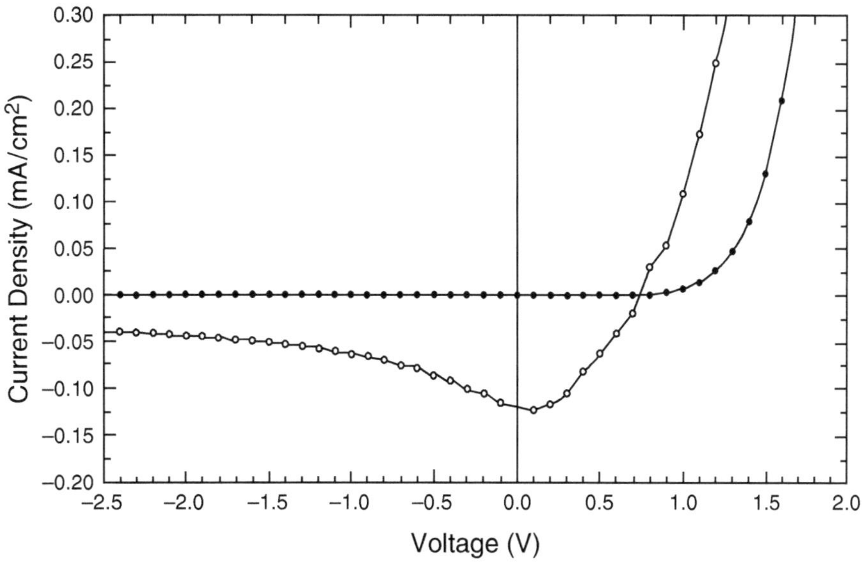

Fig. 6.2 I–V characteristics of the ITO/P3OT-SWNTs/Al device measured in the dark (*filled circles*) and under illumination (*open circles*). Adapted from [9]

which are able to split up the excitons and also create a continuous pathway for the electrons to be efficiently transported to the negative electrode.

CNTs were also added into the polymer P3HT as the hole acceptor to improve the performance of the polymer–buckminsterfullerene (C_{60}) device [10]. In the prepared ITO/P3HT-CNT/C_{60}/Al cell, the C_{60} layer would act as both the electron acceptor and the electron transporting layer when excitons are dissociated at the polymer–C_{60} interfaces. Since the added multiwalled carbon nanotubes (MWCNTs) do not contribute in the photogeneration process, the nanotubes in a polymer layer would act as exciton-dissociating centers. The CNTs in the polymer–CNT composites would also expectedly provide percolating channels for hole transport to the electrode. Figure 6.3 shows the compare of the device performance for the ITO/P3HT/C_{60}/Al and ITO/P3HT-CNT/C_{60}/Al structures with 2% and 20% MWCNT. An increase of open-circuit voltage and short-circuit current in the polymer–CNT/C_{60} devices compared with the polymer/C_{60} ones is observed. A V_{OC} of 0.7–0.8 V is obtained for the ITO/P3HT-CNT/C_{60}/Al device, which is close to the difference between the work functions of CNT (5.1 eV) and Al (4.3 eV).

In addition, CNTs could also be used as the conducting scaffolds to support semiconducting quantum dots (QDs) [11] or dyes [14] attached onto them in the polymer solar cell. Figure 6.4 shows the image of a CdSe-SWCNT complex. The incorporation of this QD-SWCNT or dye-SWCNT complexes into the polymer composite solar cells could result in higher cell efficiencies.

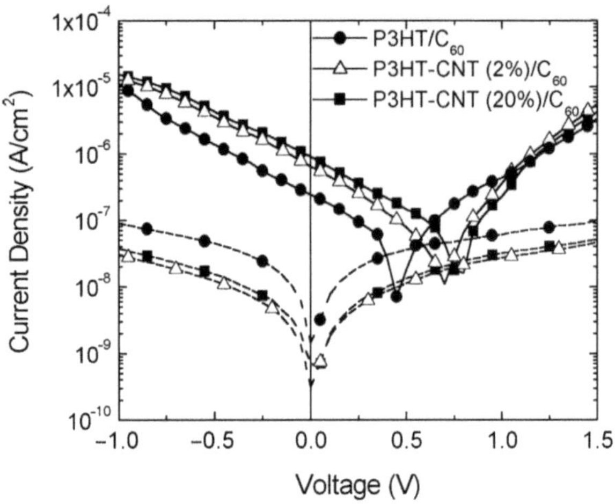

Fig. 6.3 I–V characteristics of the ITO/P3HT/C_{60}/Al and ITO/P3HT–CNT/C_{60}/Al device with 2% and 20% CNT in dark (*broken lines*) and under white light illumination (*solid lines*). Images adapted from [10]

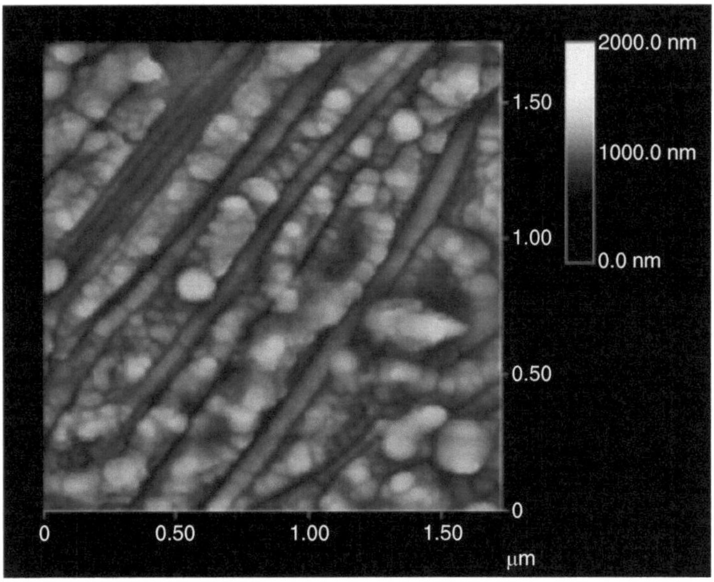

Fig. 6.4 AFM image of CdSe-SWNT complex. Adapted from [11]

6.2.2 Using Carbon Nanotubes as Transparent Electrodes

The CNTs can also be used as the transparent electrodes for collecting charge carriers in organic solar cell applications [15–18]. Figure 6.5a shows the structure of a CNT/PPV/Al photovoltaic device, in which the MWCNT film is spin-coated as the

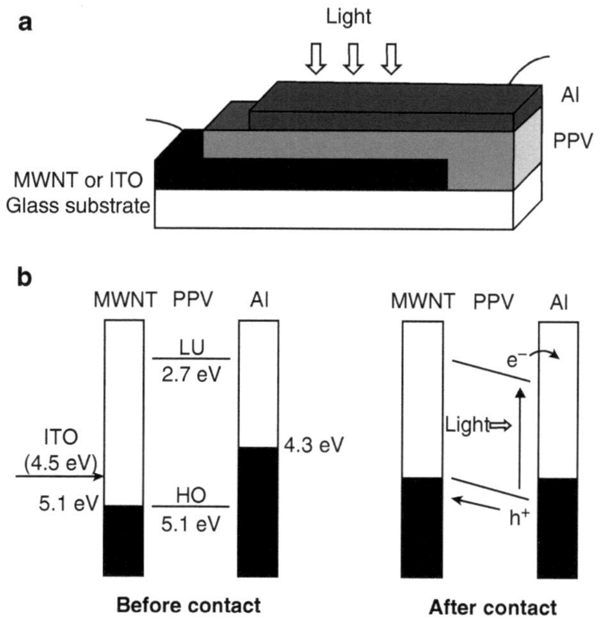

Fig. 6.5 (**a**) Structure of the MWCNT/PPV/Al photovoltaic device. Visible light is shone through a semi-transparent Al electrode. The film thickness of PPV and MWCNT is 210 and 140 nm, respectively. (**b**) Energy diagram of the MWCNT/PPV/Al device before and after contact. Adapted from [15]

hole-collecting electrode to replace the traditional indium tin oxide (ITO) glass electrode [15]. The MWCNT layer was used as the electrode due to its relatively high conductivity (8 S/cm at room temperature) [19] and high work function (5.1 eV) [20]. The energy diagram of the MWCNT/PPV/Al device before and after contact is shown in Fig. 6.5b. As shown in Fig. 6.5b, a lower barrier to hole-collection for the MWCNTs can be acquired compared with ITO. The scanning electron microscope (SEM) observation also showed that a very high contact area between poly(*p*-phenylene vinylene) (PPV) and MWCNT layers could be achieved. These factors are advantageous to the collection of the holes and cause I_{SC} to be effectively increased. The maximum V_{OC} of the photovoltaic device is expected to be decided by the difference in the work functions of the two electrodes. Since the work function of ITO (4.5 eV [21]) is smaller than that of the MWCNT (5.1 eV [20]), a higher V_{OC} can be expected for an MWCNT/PPV/Al device than for an ITO/PPV/Al device. At the illumination of the light with a wavelength of 485 nm and an intensity of 37 mW cm^{-2}, a V_{OC} and a I_{SC} of 0.90 V and 0.56 mA cm^{-2} are obtained for the device respectively. The FF and power conversion efficiency were calculated to be 0.23 and 0.081% respectively.

SWCNT networks were also formed on plastic by a transfer-printing method [22] to use as transparent electrodes for efficient, flexible polymer–fullerene bulk–heterojunction solar cells [16]. The transfer-printing method can produce

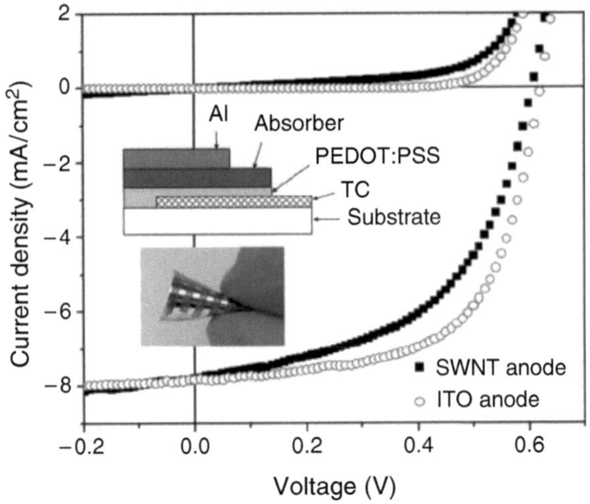

Fig. 6.6 Current density–voltage characteristics of P3HT:PCBM devices under AM1.5G conditions using ITO on glass (*open circles*) and flexible SWCNTs on PET (*solid squares*) as the anodes, respectively. *Inset*: Schematic of device and photograph of the highly flexible cell using SWCNTs on PET. Adapted from [16]

homogeneous, relatively smooth films, enabling the fabrication of conjugated polymer-based bulk–heterojunction solar cells without the use of layers thicker than optimal. The SWCNT films fabricated by this way have a transparency of 85% in the visible and a sheet resistance of 200 Ω/Υ. Figure 6.6 shows current density–voltage characteristics of the fabricated devices with the PET/SWCNTs/PEDOT:PSS/P3HT:PCBM/Al structure and the glass/ITO/PEDOT:PSS/P3HT:PCBM/Al structure respectively. As seen in Fig. 6.6, a V_{OC} and an I_{SC} of 605 mV and 7.8 mA cm^{-2} are obtained by the PET/SWCNTs/PEDOT:PSS/P3HT:PCBM/Al device. It was shown that the devices fabricated using SWCNT networks operate almost identically to those for ITO-coated glass with the exception of the FF, which is likely attributable to the relatively high series resistance for the SWCNT film. The achieved power efficiency of 2.5% (AM1.5G) made with SWCNTs/PET approaches that of the control device made with ITO/glass (3%). A main merit of the device with SWCNT/plastic electrodes is that its flexibility is far superior to devices using ITO-coated on the same flexible substrate material.

6.3 Fabrication of Nanowelded Carbon-Nanotube Solar Photovoltaic Microcells

High aspect ratios and the large surface area for nanotubes are advantageous to exciton dissociation, charge carrier transport and collection, which can improve the power conversion efficiency of the photovoltaic devices. However, the D–A type solar cell based on the composite of conjugated polymers and SWCNTs still shows a low power conversion efficiency (<0.1%) at present [9–12, 14]. The low hole mobility for polymers and the difficult acquirement of the good dispersion for nanotubes in polymers have limited the enhancement of the device performance. In addition, the solar cells with CNT films as transparent electrodes have also presented no significant improvement of efficiency compared with the cells fabricated by ITO glass [16–18]. In those devices, nanotubes do not participate in the photogeneration process, which is disadvantageous to illustrate clearly the application potential of SWCNTs in the photovoltaic field. In the following description, we will introduce a solar photovoltaic cell directly using SWCNTs as the energy conversion material, which will fully utilize the excellent photoelectric properties of SWCNT itself and show an improved device performance.

Figure 6.7 shows the diagrammatic sketch of this PV device. Parallel palladium (Pd) and aluminum (Al) electrodes, which have high and low work function (Φ) of 5.1 and 4.1 eV respectively, were patterned on the silicon (Si) wafer coated with a 500-nm thermally oxidized layer by the standard UV lithography and lift-off process. The electrodes fabricated were 50 μm long, 40 μm wide, 150 nm thick, and have a gap of 500 nm. The Si substrate with sputtered Al electrode was used as the back gate to tune the band bending of the device. The SWCNTs with an average

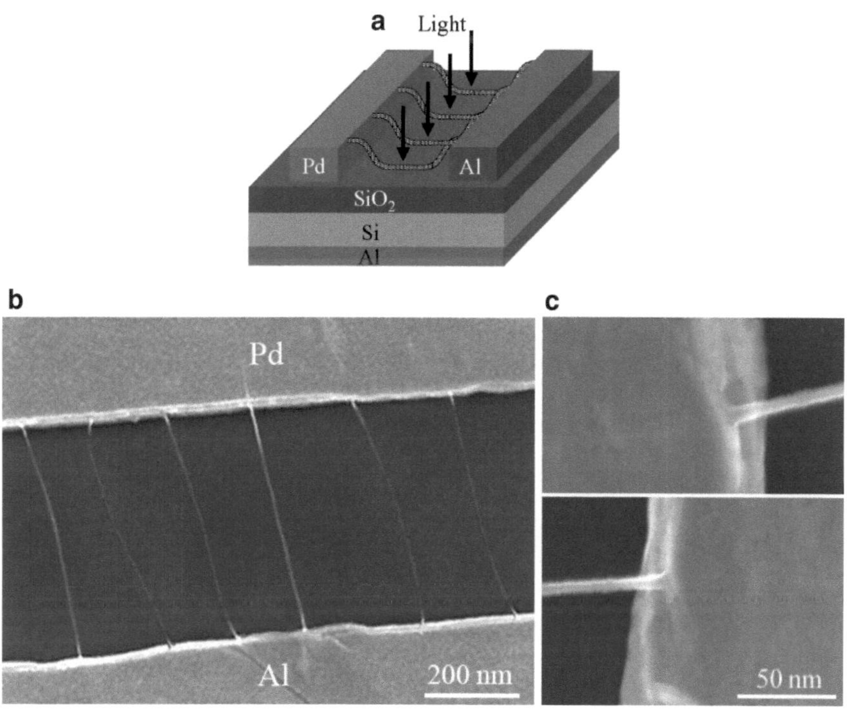

Fig. 6.7 Structure of the SWCNT PV microcells based on nanowelded high work function metal/nanotube/low work function metal junctions. (**a**) The schematic diagram of SWCNT PV cells. (**b**) SEM image of SWCNT array bridging the Pd and Al electrodes. (**c**) SEM images of the SWCNT ends nanowelded into the metal electrodes

diameter of 0.9 nm, the work function and the band gap of which are $\Phi \sim 4.5$ eV and $E_g \sim 1.1$ eV respectively [23], were aligned onto two electrodes by the AC dielectrophoresis method [24] to form the dispersedly directed array of SWCNT bundles. Afterward, SWCNTs were welded into the electrodes by the ultrasonic nanowelding technique introduced in the Chap. 4 so as to form the firm and low-resistance SWCNT/metal contacts [25, 26]. The SEM image of the SWCNT bundle array bridging the electrodes and the welded SWCNT ends is shown in Fig. 6.7b, c. A stage-by-stage increased negative bias was then applied to the Pd electrode to selectively destroy the metallic ones in the aligned SWCNTs [27]. In this process, due to the fact that only a very small reverse-biased current passed through them, the semiconducting SWCNTs would retain not to be destroyed. To avoid the effects caused by the photo-desorption of oxygen [28, 29] in the illumination test, the fabricated devices were protected by a thin ($\sim$10 nm thick) CVD SiO_2 film [7].

The as-fabricated devices were tested under the illumination of the solar-simulated light with different incident power densities, the spectra of which were the same as those of AM1.5G and which was produced by a calibrated xenon lamp. The direction of the incident light was vertical to the surface of the Si substrate.

6.4 Characteristics of Nanowelded Carbon-Nanotube Solar Photovoltaic Microcells

6.4.1 Photovoltaic Effect of the Device

In dark, the device shows rectifying I–V characteristics at $V_G = 0$ (Fig. 6.8a) due to the formation of Schottky barriers at metal–nanotube contacts [30–34]. The early studies have shown when metal Pd is contacted to the SWCNT, its Fermi level would

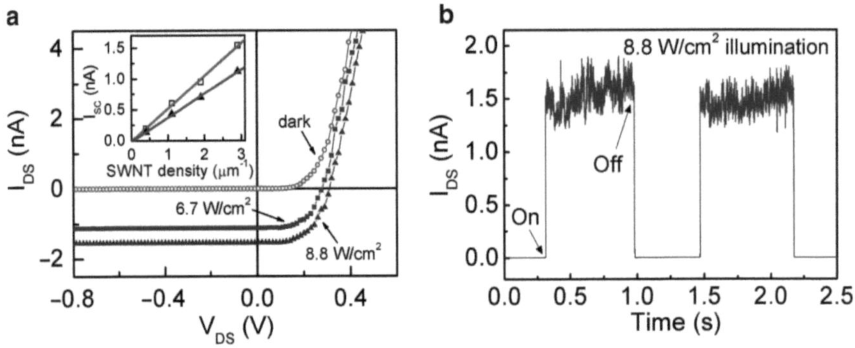

Fig. 6.8 (**a**) I–V characteristics measured in dark and under illumination at $V_G = 0$, showing the PV effect of the devices. The inset displays the dependency of I_{SC} on SWCNT density (= SWCNT number/electrode length) in the devices; the lines with triangle and square represent the data obtained at the illumination intensity of 6.7 and 8.8 W cm^{-2}, respectively. (**b**) The time-resolved trace of the current in current in the device with the light chopped, taken with zero drain bias and a gate voltage of $V_G = 0$

be located at or even slightly lower than the top of the valence band of SWCNTs [30, 31], and when metal Al is contacted to the SWCNT, its Fermi level would be close to the bottom of the conduction band of SWCNTs [30]. Thus, the device structure is analogous to a p–i–n junction diode, exhibiting the rectifying I–V characteristics. The result is consistent with those reported by asymmetrically contacted nanotubes in previous literatures [34–37]. The obtained typical I–V characteristics in dark can be well fitted by the standard diode equation, giving a reverse saturation current of $I_S \sim 5$ pA, an ideality factor of $n \sim 1.9$, and a series resistance of $R_S \sim 271\,\Omega$.

Under the illumination of sunlight, this device shows PV effects. As shown in Fig. 6.8a, at an illumination power density of 6.7 W cm^{-2}, the short-circuit current (I_{SC}) and open-circuit voltage (V_{OC}) of the device were 1.12 nA and 0.28 V at $V_G = 0$ respectively. When the illumination power density of 8.8 W cm^{-2} was applied, I_{SC} and V_{OC} increased to 1.54 nA and 0.31 V respectively. The obtainment of the PV effect is due to a strong electric field along the nanotube being formed by contacting SWCNTs to two asymmetrical metal electrodes, which efficiently separates the photogenerated e–h pairs. Figure 6.8b shows the relation of current and time for the devices at zero drain bias when switching the incident light. As shown in Fig. 6.8b, the photocurrent could be produced straightway as the incident light was turned on, with a response time of ~90 ms.

6.4.2 Physical Mechanism of the Device

The operation mechanism of this SWCNT PV cell can be illustrated by the energy band diagram of the device in the initial equilibrium state (Fig. 6.9a). The contact of SWCNTs to metal electrodes causes the bending of the energy bands in SWCNTs and the formation of Schottky barriers at the metal–nanotube contacts. For no Fermi level pinning, when SWCNTs are contacted to the high and low work-function metal electrodes, p-type and n-type Schottky barriers are formed respectively in the two ends of SWCNTs. For that the gate oxide thickness is on the order of (equal to) the SWCNT channel length in our devices, the bending of the SWCNT band would extend into the device middle because of the weak coupling of gate to channel [38–40]. Previous studies have also indicated that the bending of the band could

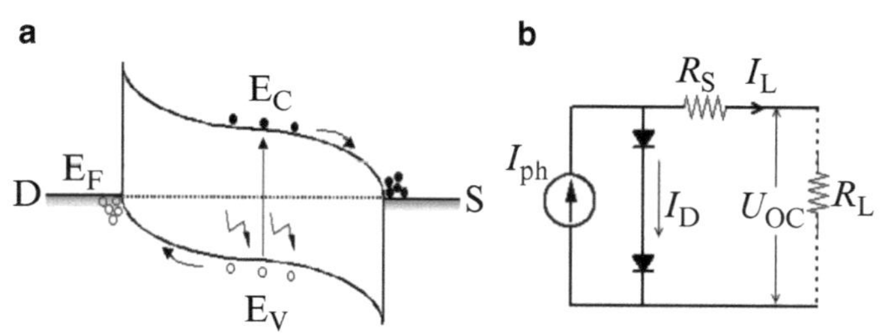

Fig. 6.9 Analysis of the mechanism for the SWCNT PV cells. (**a**) The energy band diagram of SWCNTs bridging two dissimilar metal electrodes, showing the currents from photo-generated e–h pairs. (**b**) The equivalent circuit of the cells

extend to a distance of microns or further when SWCNTs (lightly doped) are connected to the metal [40]. Thus, a strong built-in electric field spanning the whole SWCNT channel could be formed to separate the carriers, which is ideal for the efficient carrier separation. Figure 6.9b shows the equivalent circuit of this SWCNT PV cell.

6.4.3 Power Conversion Efficiency of the Device

η of the PV device can be obtained from the fourth-quadrant I–V curve, which can be expressed as $\eta = (V_m I_m)/P_{in} = (FF \cdot V_{OC} I_{SC})/P_{in}$, where FF is the fill factor, V_m and I_m are the output voltage and current with which the output power $(= V_m I_m)$ is maximum, and P_{in} is the incident light power. P_{in} is estimated by considering the total area of the SWCNT cross-sections perpendicular to the incident direction as the actual effective active area of the devices, which was obtained by carefully measuring the actual diameter and spanning length of each SWCNTs in the device with the atomic force microscope and SEM. The measure result shows that an average diameter of 5.2 nm for the SWCNTs was obtained, which is due to the fact that SWCNTs tend to form the bundles. Correspondingly, a P_{in} of ~33.4 nW is estimated for the device with 146 SWCNT bundles bridging the electrodes under the illumination intensity of 8.8 W cm^{-2}. Thereby, a photoresponsivity (I_{SC}/P_{in}) and η of the device can be deduced to be ~46.7 mA W^{-1} and ~0.80% respectively. Here, it is to expect reflecting the photoelectricity conversion characteristics of SWCNT active region itself in the device better that using the SWCNT area but not the device substrate area to calculate P_{in} and η. Our estimation of η is reasonable, considering that the methods to increase the nanotube coverage maybe be used in the fabrication of future devices. Note that the obtained η is high virtually for that the thin monolayer SWCNTs were used to build the devices and only a small fraction of incident power was absorbed by the SWCNTs.

To further explore the device, the dependency of the cell performance parameters on the incident power density (P_0) was studied. It is seen from Fig. 6.10 that both

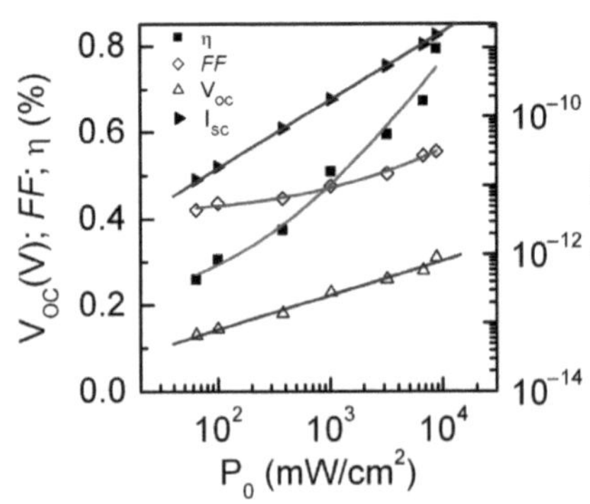

Fig. 6.10 Key cell performance parameters as a function of the incident power density

log (I_{SC}) and V_{OC} increase logarithmically with P_0, and FF increases from 0.42 to 0.56 with increased P_0 in the measure range. As a result, η increases from ~0.26% to ~0.80% when the incident power increases from 64 mW cm^{-2} to 8.8 W cm^{-2}. At $P_0 = 100$ mW cm^{-2} (1 sun intensity), the cell shows a η of ~0.31%. The research result shows that increasing the illumination power density can effectively enhance the cell efficiency.

6.5 Simulation of Carbon-Nanotube Solar Photovoltaic Microcells

Only a small fraction of incident power can be absorbed to contribute to the optoelectronic conversion, because this device is constructed by the monolayer SWCNTs. To estimate the actual power absorption of SWCNTs under the solar spectrum illumination and the internal η of SWCNTs in this device, the high-frequency electromagnetic finite-element software HFSS can be used to make a simulation [41]. In the simulation, each SWCNT bundle is modeled as a dielectric cylinder and the periodic boundary condition is used to deal with the array structure. The detailed simulation is introduced as follows.

At first, the dielectric function of SWCNTs is calculated with the method in [42]. The chiral vector (9,4) is presumed for our SWCNTs in the calculation. Figures 6.11a, b are respectively the calculated imaginary part of the SWCNTs' dielectric function when the electric field of the incident light is parallel and perpendicular to the SWCNT axis. The real part of the dielectric function can be derived from the imaginary part according to the Kramers–Kronig relation [43].

The system's response to the incident optical field is then simulated by the finite-element software HFSS. Considering that the solar light is randomly polarized in the plane vertical to the incident direction, half of the incident power can be assigned along the SWCNT's axis and another half is assigned perpendicular to its axis. For these two cases, we calculated the monochromatic absorbed powers at various incident phonon energies, which range from 0.5 to 4 eV with a step length of 0.01 eV. This range is enough to cover the solar spectrum. Figure 6.12 shows the simulated optical field distribution under the specific monochromatic illuminations for the electric field polarization of the light parallel and perpendicular to the tube axis. The absorbed powers per SWCNT bundle as a function of the photon energy for these two polarization cases are shown in Fig. 6.13a, b, respectively. The absorbed powers in the figure are normalized to the monochromatic incident power density of 1 W m^{-2}. It is indicated that the absorbed power for the electric field polarization of the light perpendicular to tube axis is much smaller (about a magnitude smaller) than that for the electric field polarization of the light parallel to tube axis, which results from the smaller dielectric function and the depolarization effect for the former case [44].

By integrating the power absorption spectrum weighted by the power density component of the used solar-simulated spectrum under different photon energies,

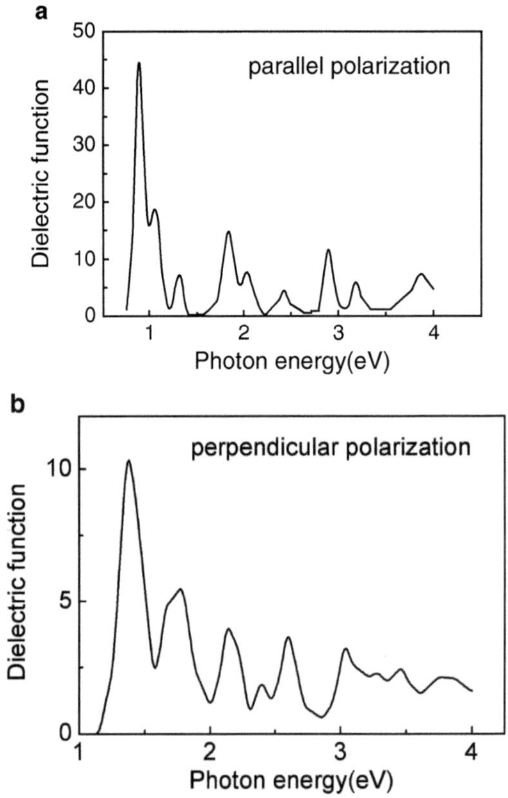

Fig. 6.11 The imaginary part of dielectric function of the SWCNT (9,4) for the electric field of the light parallel (**a**) and perpendicular (**b**) to the SWCNT axis, respectively

we can obtain the total absorbed power under the solar-simulated light illumination, which can be expressed as

$$P = \int \frac{\mathrm{d}P_S}{\mathrm{d}(\hbar\omega)} \times P_a(\hbar\omega)\mathrm{d}(\hbar\omega)$$

where $\mathrm{d}P_S/\mathrm{d}(\hbar\omega)$ is the incident power density component distribution with the photon energy and $P_a(\hbar\omega)$ is the absorbed power per unit monochromatic incident power density. The calculated total absorbed power per SWCNT bundle is about 1.41×10^{-11} W for the incident power density of 8.8 W cm^{-2}. For the 146 SWCNT bundle array, a total absorbed power of 2.06 nW is obtained, while when the incident power density is 100 mW cm^{-2}, a total absorbed power of about 0.023 nW for 146 SWCNT bundle array is calculated. Similarly, the total absorbed powers under other incident power densities can also be calculated.

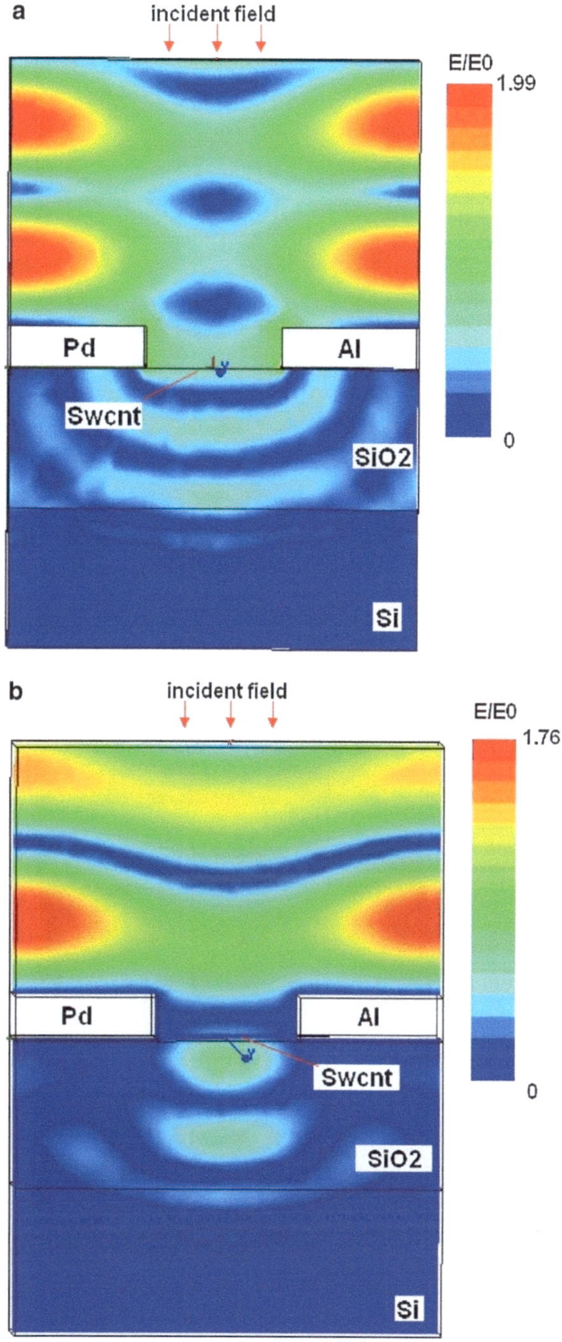

Fig. 6.12 The optical field distribution under monochromatic illumination of 1.4 eV for the parallel (**a**) and perpendicular (**b**) polarization case, and of 2.5 eV for the parallel (**c**) and perpendicular (**d**) polarization case, respectively. E0 is the amplitude of the incident optical field

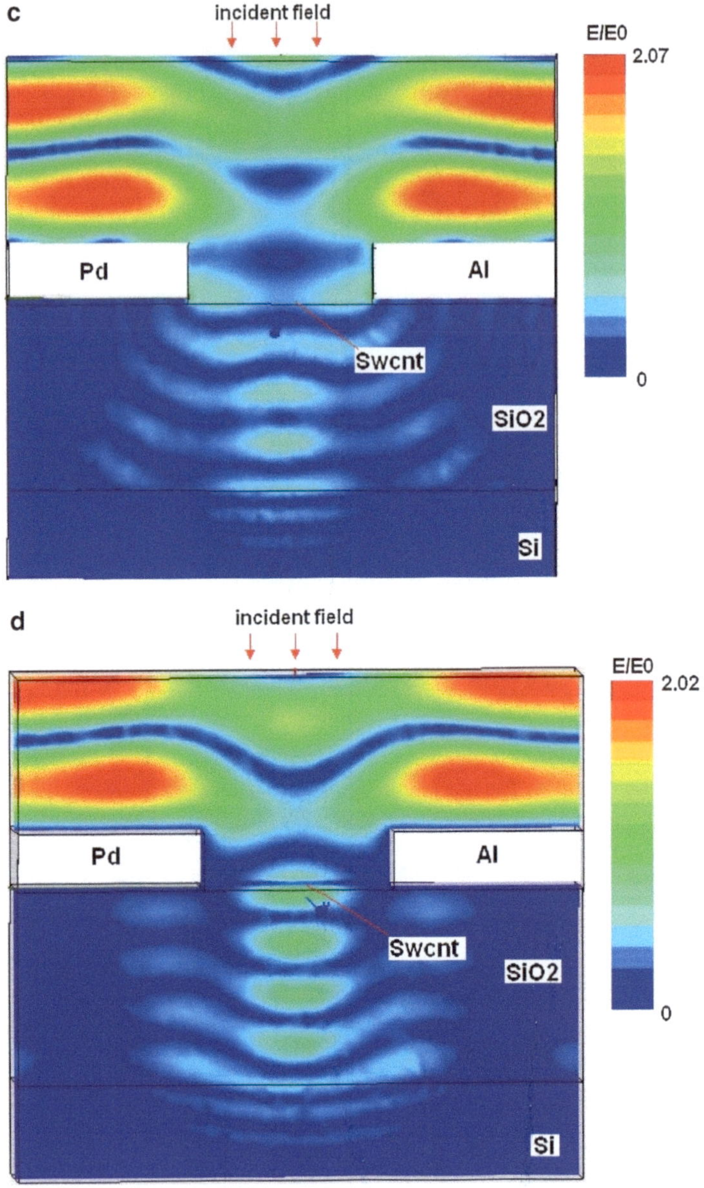

Fig. 6.12 (continued)

Based on the calculated actual absorbed solar power for the SWCNTs in the device, an internal η of 12.6% and 5.1% can be estimated correspondingly for SWCNTs under the illumination intensity of 8.8 and 100 mW cm^{-2} respectively.

To illustrate the acquirement of high internal η, a key reason would be the nicer designed device structure, which can cause a strong built-in electric field and a wide

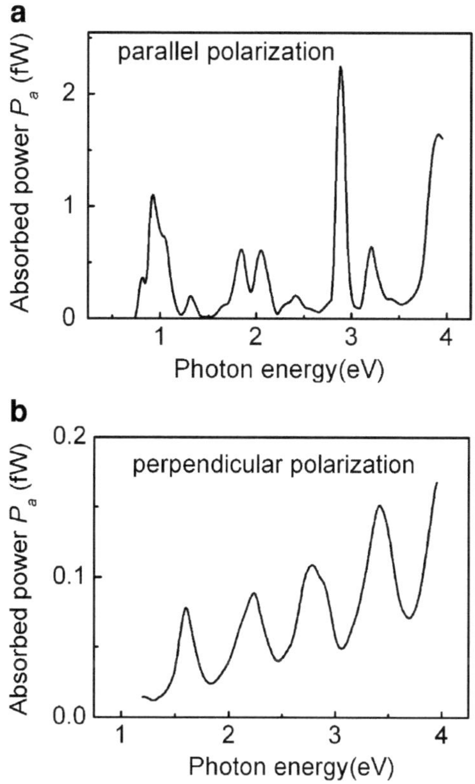

Fig. 6.13 The power absorption spectra per SWCNT bundle as a function of the photon energy for the parallel (**a**) and perpendicular (**b**) polarization case. The absorbed power under different photon energy is normalized to per unit monochromatic incident power density

photoactive region formed in the device. For the present device, the photoactive region spans the whole SWCNT between two metal electrodes, while for the SWCNT p–n junction the photoactive region is only near the narrow p–n junction [45]. The other important factor is that the series resistance of PV cells, which mainly arises from the nanotube–metal contact resistance [34], was significantly decreased by applying the ultrasonic nanowelding technology. Our previous studies had demonstrated that ultrasonic nanowelding treatment can decrease the contact resistance by three to four orders of magnitude [25].

6.6 Conclusion

SWCNTs had been used as the acceptor and transport path of the charge carriers or the transparent electrodes in organic solar photovoltaic cells. However, the η obtained by these cells was still low and showed no significant improvement compared with the cells fabricated by ITO glass. In these devices, nanotubes do not participate in the photogeneration process, but are only utilized for their high aspect ratios and large surface area.

To further explore the application potential of SWCNTs in the photovoltaic field, a novel solar photovoltaic cell directly using SWCNTs as the energy conversion material has been studied. In this SWCNT PV device, directed monolayer SWCNTs have been ultrasonically nanowelded onto two asymmetrical metal electrodes with high and low work function causing a strong built-in electric field in SWCNTs to separate photogenerated electron-hole pairs efficiently. In dark, the devices exhibited rectifying I–V characteristics. Under *solar-light* illumination, the devices showed PV effects. A power conversion efficiency of 0.80% and 0.31% was achieved respectively at an illumination intensity of 8.8 and 100 mW cm^{-2} for monolayer SWCNT cells. Correspondingly, an internal power conversion efficiency of 12.6% and 5.1% could be estimated for SWCNTs in the device based on the actual absorbed incident power of SWCNTs calculated by the device simulation.

The high η obtained by this PV cell can be attributed to the combination of the favorable device structure design and the significantly decreased series resistance achieved by the ultrasonic nanowelding technique. Further improvements in η can be expected by choosing SWCNTs with more suitable band gap, adopting the multilayered different diameters of SWCNT arrays, and further decreasing the channel length and the series resistance of the cells. The research results show a great application potential for SWCNTs in the PV field.

References

1. M.E. Green, *Solar Cells: Operating Principles, Technology, and Applications* (Prentice–Hall, Englewood Cliffs, NJ, 1982)
2. G. Yu, J. Gao, J. Hummelen, F. Wudl, A.J. Heeger, *Science* 270, 1789 (1995)
3. S.G. Bailey, S.L. Castro, R.P. Raffaelle, S. Fahey, T. Gennett, P. Tin, in *Proceedings of 3rd World Conference on Photovoltaic Energy Conversion*, vol. 3, p. 2690 (Osaka, Japan, 2003)
4. J. Ando, J. Phys. Soc. Jpn. **66**, 1066 (1997)
5. T.G. Pedersen, Phys. Rev. B. **67**, 073401 (2003)
6. R. Saito, G. Dresselhaus, M.S. Dresselhaus, *Physical Properties of Carbon Nanotubes* (Imperial College, London, UK, 1998)
7. C.X. Chen, Y. Lu, E.S. Kong, Y.F. Zhang, S.T. Lee, Small **4**, 1313 (2008)
8. S. Barazzouk, S. Hotchandani, K. Vinodgopal, P.V. Kamat, J. Phys. Chem. B **108**, 17015 (2004)
9. E. Kymakis, G.A.J. Amaratunga, Appl. Phys. Lett. **80**, 112 (2002)
10. B. Pradhan, S.K. Batabyal, A.J. Pal, Appl. Phys. Lett. **88**, 093106 (2006)
11. B.J. Landi, R.P. Raffaelle, S.L. Castro, S.G. Bailey, Prog. PhotoVoltaics: Res. Appl. **13**, 165 (2005)
12. C.J. Brabec, B.N.S. Sariciftci, J.C. Hummelen, Adv. Funct. Mater. **11**, 15 (2001)
13. M. Cinke, J. Li, B. Chen, A. Cassell, L. Delzeit, J. Han, M. Meyyappan, Chem. Phys. Lett. **365**, 69 (2002)
14. S. Bhattacharyya, E. Kymakis, G.A. Amaratunga, J. Chem. Mater. **16**, 4819 (2004)
15. H. Ago, K. Petritsch, M.S.P. Shaffer, A.H. Windle, R.H. Friend, Adv. Mater. **11**, 1281 (1999)
16. M.W. Rowell, M.A. Topinka, M.D. McGehee, H.J. Prall, G. Dennler, N.S. Sariciftci, L.B. Hu, G. Gruner, Appl. Phys. Lett. **88**, 233506 (2006)

17. J. van de Lagemaat, T.M. Barnes, G. Rumbles, S.E. Shaheen, T.J. Coutts, C. Weeks, I. Levitsky, J. Peltola, P. Glatkowski, Appl. Phys. Lett. **88**, 233503 (2006)
18. A.J. Miller, R.A. Hatton, G.Y. Chen, S.R.P. Silva, Appl. Phys. Lett. **90**, 023105 (2007)
19. H. Ago, M.S.P. Shaffer, D.S. Ginger, A.H. Windle, R.H. Friend, Phys. Rev. B **61**, 2286 (2000)
20. H. Ago, T. Kugler, F. Cacialli, W.R. Salaneck, M.S.P. Shaffer, A.H. Windle, R.H. Friend, J. Phys. Chem. B **103**, 8116 (1999)
21. J.S. Kim, M. Gransröm, R.H. Friend, N. Johansson, W.R. Salaneck, R. Daik, W.J. Feast, F. Cacialli, J. Appl. Phys. **84**, 6859 (1998)
22. Y. Zhou, L. Hu, G. Gruner, Appl. Phys. Lett. **88**, 123109 (2006)
23. S.J. Tans, A.R.M. Verschueren, C. Dekker, Nature **393**, 49 (1998)
24. C.X. Chen, Y.F. Zhang, J. Phys. D: Appl. Phys. **39**, 172 (2006)
25. C.X. Chen, L.J. Yan, E.S.W. Kong, Y.F. Zhang, Nanotechnology **17**, 2192 (2006)
26. C.X. Chen, D. Xu, E.S.W. Kong, Y.F. Zhang, IEEE Electron Device Lett. **27**, 852 (2006)
27. P.G. Collins, M.S. Arnold, P. Avouris, Science **292**, 706 (2001)
28. R.J. Chen, N.R. Franklin, J. Kong, J. Cao, T.W. Tombler, Y. Zhang, H. Dai, Appl. Phys. Lett. **79**, 2258 (2001)
29. M. Shim, G.P. Siddons, Appl. Phys. Lett. **83**, 3564 (2003)
30. A. Javey, Q. Wang, W. Kim, H. Dai, Proceeding IEDM 741 (2003)
31. A. Javey, J. Guo, Q. Wang, M. Lundstrom, H. Dai, Nature **424**, 654 (2003)
32. J.J. Palacios, P. Tarakeshwar, D.M. Kim, Phys. Rev. B **77**, 113403 (2008)
33. S. Heinze, J. Tersoff, R. Martel, V. Derycke, J. Appenzeller, P. Avouris, Phys. Rev. Lett. **89**, 106801 (2002)
34. H.M. Manohara, E.W. Wong, E. Schlecht, B.D. Hunt, P.H. Siegel, Nano Lett. **5**, 1469 (2005)
35. C. Lu, L. An, Q. Fu, J. Liu, H. Zhang, J. Murduck, Appl. Phys. Lett. **88**, 133501 (2006)
36. H. Li, Q. Zhang, N. Marzari, Nano Lett. **8**, 64 (2008)
37. E. Cobas, M.S. Fuhrer, Appl. Phys. Lett. **93**, 043120 (2008)
38. M. Freitag, Y. Martin, J.A. Misewich, R. Martel, P. Avouris, Nano Lett. **3**, 1067 (2003)
39. N. Neophytou, J. Guo, M. Lundstrom, IEEE Trans. Nanotechnol. **5**, 385 (2006)
40. F. Léonard, J. Tersoff, Phys. Rev. Lett. **83**, 5174 (1999)
41. C.X. Chen, L. Yang, Y. Lu, G.B. Xiao, Y.F. Zhang, IEEE Trans. Nanotechnol. (2008). Published online
42. G.L. Zhao, D. Bagayoko, L. Yang, J. Appl. Phys. **99**, 114311 (2006)
43. P.Y. Yu, M. Cardona, *Fundamentals of Semiconductors* (Springer-Verlag, Berlin, 2001)
44. S. Tasaki, K. Maekawa, T. Yamabe, Phys. Rev. B **57**, 9301 (1998)
45. D.A. Stewart, F. Léonard, Nano Lett. **5**, 219 (2005)

Index

MIX
Papier aus verantwortungsvollen Quellen
Paper from responsible sources
FSC® C105338

If you have any concerns about our products,
you can contact us on
ProductSafety@springernature.com

In case Publisher is established outside the EU,
the EU authorized representative is:
Springer Nature Customer Service Center GmbH
Europaplatz 3, 69115 Heidelberg, Germany

Printed by Libri Plureos GmbH
in Hamburg, Germany